SpringerBriefs in Molecular Science

More information about this series at http://www.springer.com/series/8898

Pradip K. Dutta · Vinod Kumar

Optically Active Polymers

A Systematic Study on Syntheses and Properties

 Springer

Pradip K. Dutta
Department of Chemistry
Motilal Nehru National Institute
  of Technology
Allahabad, Uttar Pradesh
India

Vinod Kumar
Department of Chemistry
Motilal Nehru National Institute
  of Technology
Allahabad, Uttar Pradesh
India

ISSN 2191-5407   ISSN 2191-5415   (electronic)
SpringerBriefs in Molecular Science
ISBN 978-981-10-2605-8   ISBN 978-981-10-2606-5   (eBook)
DOI 10.1007/978-981-10-2606-5

Library of Congress Control Number: 2016950887

Printed on acid-free paper

This Springer imprint is published by Springer Nature
The registered company is Springer Nature Singapore Pte Ltd.
The registered company address is: 152 Beach Road, #22-06/08 Gateway East, Singapore 189721, Singapore

Preface

Recently, the ranking of world institutions has been the front page news to all academicians, researchers, students and industrialists. Nowadays, innovation and constructive research are required in every field to survive globally. As a result, research has gained remarkable momentum in every aspect of science, engineering, technology and social sciences. All nations of the world are focusing on need-based research for further development. Research is the sense of mind, imagination and creation of a person which enumerates the results or shape to the reality. In true sense, research has multi-dimensional phases and approaches. Like in general, a physicist should prefer to work on physics route, a chemist on chemistry and so on. But there are many areas which are composite in nature, where it needs multi-talent expertise and it is only possible from researchers of various disciplines. Optically active polymers are such materials where a person of different fields of research can work. To carry out advanced studies or present need-based research, it is mandatory to understand the material first. Optically active materials can rotate into a plane of polarization and its activity originates from the presence of chiral elements in a polymer such as chiral centres or chiral axes due to long-range conformational order in a macromolecule. Surprisingly, most naturally occurring macromolecules possess the ability to organize to more complex high structure rather than single one and manifest their functions. The problems of charged and reactive polymers are correlated to optically active compounds and it shows an inherent property of both ordinary macromolecules and large range of synthetic polymers. It is true that the properties of a material depend on the synthesis methods. Hence to know all such characteristics, a good researcher should begin the study of optically active polymers in a systematic manner like from synthesis to property. The present brief entitled "Optically active polymers: A systematic study on syntheses and properties" is an attempt to provide the researchers of various backgrounds with a thorough understanding and knowledge of the optically active polymers with a special emphasis on syntheses and properties. We sincerely thank all learned researchers whose works have been the major source motivation for the preparation of this brief. We are also thankful to the editorial board and publisher for accepting our proposal.

Allahabad, India   Pradip K. Dutta
May 2016   Vinod Kumar

Acknowledgement

We gratefully acknowledge the research environment support provided to us in the Department of Chemistry of the Institute during the preparation of the manuscript. The authors also express their gratitude to various authors of different literature and researchers whose works have been the essential source and support to prepare the manuscript in its present form.

About the Book

The brief presents a systematic study of synthesis of optically active polymers. It discusses in detail about the syntheses of three different types of optically active polymers from helical polymers, dendronized polymers and other types of polymeric compounds. The brief also explains the syntheses of optically active azoaromatic and carbazole containing azoaromatic polymers and copolymers; optically active benzodithiophene; and optically active porphyrin derivatives. The final chapter of the brief discusses different properties of optically active polymers such as nonlinear optical properties, chiroptical properties, vapochromic behavior, absorption and emission properties, fabrication and photochromic properties. The intrinsic details of different properties of optically active polymers will be useful for researchers and industry personnel, who are actively engaged in application oriented investigations.

Contents

About the Authors

Dr. Pradip K. Dutta is Professor and Former Head, Department of Chemistry, Motilal Nehru National Institute of Technology, Allahabad, India and founder editor of Asian Chitin Journal, an international journal since 2005. He obtained his M.Sc. (1987) and Ph.D. (1993) from IIT Kharagpur. His specialization is in Physical/Polymer Chemistry and research interests include synthesis, modification, physical, chemical and biological properties of engineering polymers: optically active polymers, chitosans, scaffolds for biomedical applications, nanocomposites preparation and application to tissue engineering, drug delivery, food prevention and wound management. He has more than 200 research publications and supervised 13 Ph.D. students, 19 M.Tech./M.Phil./M.Pharm dissertation, 16 M.Sc. dissertations and 2 patents. At present, 4 Ph.D. students are working under him. He is the author/editor of many books/chapters/course materials under continuing education programme (AICTE) and reviewer of many national and international journals. He has handled about a dozen of research projects as principal investigator. He has visited many foreign countries like USA, UK, China, Japan, South Korea, Switzerland, Turkey for academic purposes. He has vast experience in academic/research/administration. He is the recipient of many national/international fellowships. He was honoured by Royal Society of Chemistry, UK as fellow (FRSC) in 2007. e-mail: pkd@mnnit.ac.in

Dr. Vinod Kumar is born on 13 January 1984 in Gothua Jagipur, a village in the District of Sultanpur, Uttar Pradesh. His early education started in his village school. He obtained his M.Sc. degree in Chemistry in 2006 from Dr. Ram Manohar Lohia Avadh University, Faizabad and D.Phil. degree in Chemistry from University of Allahabad, Allahabad, in January 2012. His area of research was in analytical chemistry. During his research work he worked as UGC-project fellow. After obtaining D.Phil. degree he worked with Prof. Pradip K. Dutta in Department of Chemistry, MNNIT Allahabad in the field of polymer chemistry from March 2012 to May 2014. He worked as Senior Scientific Assistant in Forensic Science Laboratory, Govt. of NCT of Delhi, Rohini, Delhi from June 2014 to November 2014. After this he worked in the Department of Forensic Medicine & Technology, All India Institute of Medical Sciences (AIIMS), New Delhi as a research staff from December 2014 to January 2016. Again he started working since January 2016 in the field of polymer chemistry under the mentorship of Prof. Pradip K. Dutta in Department of Chemistry, MNNIT, Allahabad. He has published 16 research papers in national and international journals and attended many national and international conferences and presented his research work. e-mail: vinodvishwakarma1984@gmail.com

Abbreviations

AIBN	Azobisisobutyronitrile
APU	Advanced processing unit
ATRP	Atom transfer radical polymerization,
BDT	Benzo [1,2-b:4,5-b']dithiophene
BnEPhOx	(S)-4-benzyl-2-(ethynylphenyl)-oxazoline
CSPs	Chiral stationary phases
DABCO	1,4-diazabicyclo[2.2.2]octane
DIPC	N,N-diisopropyl-carbodiimide
DMF	Dimethylformamide
DMAP	Dimethylamino pyridine
DOPA	3-3(3,4-dihydroxyphenyl)-L/D-alanine
DPTS	4-(diphenylamino) piridinium 4-toluensulfonate
EtOAc	Ethyl acetate
GPC	Gel permeation chromatography
HCAPE-C	2-Hydroxy-N-ethyl-N-{4-[(4-cyanophenyl)-(3-carbazolyl)-diazene]}phenyl-amine
HECA	(4-cyano phenyl)-[3-[9-(2-hydroxyethyl)carbazolyl]]diazene
HMDS	1,1,1,3,3,3-hexamethyldisilazane
MCAPE-C	Methacryloyl -2-oxyethyl-N-ethyl -N-{4- [(4-cyanophenyl) (3-carbazoil) diazene]}-phenylamine
MEH-PPV	Poly 2-methoxy - 5-2- ethylhexyloxy- 1,4-phenylenevinylene
MMA	Methyl methacrylate
M_ns	Number-average molecular weights
NLO	Nonlinear optical
NMP	N-methylpyrrolidone
PAE	Poly(aryleneethynylene)
PATs	Poly(monoalkyl)thiophenes
PAVs	Poly(arylene-vinylenes)
PAE-PAV	Poly(arylene-ethynylene)-altpoly(arylene-vinylene)
PATEI	Poly(amide-thioester-imide)

PEIs	Poly(ester-imide)s
PEMU	Polyelectrolyte multilayer
PGA	Poly(glutamic acid)
P(L-HMPMA)	Poly[N-(L)-(1-hydroxymethyl)propylmethacrylamide]
PMDA	Pyromellitic dianhydride
PNIPAm	Poly(N-isopropylacrylamide)
PN(S)$_4$VP	Poly(N-(S)-2-methylbutyl-4-vinyl pyridininum iodide)
pnPEGMAN	p2PEGMAN or Poly[(E)-2-(2-(methyl(4-((4-nitrophenyl) diazenyl)phenyl) amino)ethoxy)ethyl methacrylate]
p3PEGMAN	Poly[(E)-2-(2-(2-(methyl(4-((4-nitrophenyl)diazenyl)phenyl) amino) ethoxy)ethoxy)ethyl methacrylate]
p4PEGMAN	Poly[(E)-2-(4-((4-nitrophenyl) diazenyl) phenyl) -5,8,11-trioxa - 2-azatridecan-13-yl methacrylate]
p6PEGMAN	Poly[(E)-2-(4-((4-nitrophenyl) diazenyl) phenyl) - 5,8,11,14,17-pentaoxa -2- azanonadecan -19-yl methacrylate]
PSS	Poly(styrene sulfonate)
PVC	Polyvinyl chloride
(R)-HAP-C	(R)-3-hydroxy-1-(4'- cyano-4-azobenzene)pyrrolidine
R- or S-PDOC	Poly[N-(R)- or (S)-3,7-dimethyloctyl-3,6-carbazole]
(R)-MAP-C	(R)-3-methacryloyloxy-1-(4'-cyano-4-azobenzene)pyrrolidine
TPP	Triphenyl phosphate
THF	Tetrahydrofuran
TMSA	Trimethylsilyl acetylene
(S)-BnLMA	(S)-N-leucine benzyl ester maleamic acid
(S)-BnLMI	(S)-N-maleoyl-L-leucine benzyl ester
(S)-(+)-DHDMBQT	3,3''''-di-n-hexyl-4',3'''-di[(S)-(+)-2-methylbutyl]-,2':5',2'':5'', 2''':5''', 2''''-quinquethiophene
(S)-(+)-DDDMBQT	3,3''''-Didodecyl-4',3'''-di[(S)-(+)-2-methylbutyl]-2,2':5',2'':5'', 2''': 5''',2''''-quinquethiophene
(S)-(+)-DMBTT	2-iodomagnesium-3-dodecylthiophene
(S)-HAP-C	(S)-3-hydroxy-1-(4'-cyano-4-azobenzene) pyrrolidine
(S)-HCAPP-C	(S)-3-Hidroxy-N-{4-[(4-cyanophenyl)-(3-carbazolyl)-diazene]-phenyl}- pyrrolidine
(S)-MAP	(S)-3-methacryloyloxy-1-(4-azobenzene)pyrrolidine
(S)-MAP-N	(S)-3-methacryloyloxy-1-(4'-nitro-4- azobenzene)pyrrolidine
(S)-MAP-C	(S)-3-methacryloyloxy-1-(4'-cyano-4-azobenzene)pyrrolidine
(S)-MCAPP-C	(S)-methacryloyl-3-oxy-N-{4-[(4 cyanophenyl)-(3-carbazoil)-diazene]-phenyl}pyrrolidine
(S)-(+)-MCPS	(S)-(+)-methacryloyl-2-oxy-N-9-phenylcarbazole-succinimide
(S)-(−)-MCPP	(S)-(−)-methacryloyl-3-oxy-N-9-phenylcarbazole-pyrrolidine
(S)-(+)-MECSI	(S)-(+)-methacryloyl-2-oxy-N-[3-(9-ethylcarbazole)]-succinimide
(S)-(−)-MECP	(S)-(−)-methacryloyl-3-oxy-N-[3-(9-ethylcarbazole)]-pyrrolidine

(S)-MLECA	(S)-(4-cyanophenyl)-[3-[9-[2-(2-methacryloyloxy-propanoyloxy) ethyl]carbazolyl]]diazene
(S)-MLMI	(S)-N-maleoyl-L-leucine methyl ester
(S)-(+)-MOSI	(S)-(+)-2-methacryloyloxy-N(4-azobenzene)-succinimide
(S)-MPAAP	*(S)*-3-methacryloyloxy-1-[4'-phenylazo-(4-azobenzene)]-pyrrolidine
(S)-MPAAP-C	*(S)*-3-methacryloyloxy-1-[4'-cyanophenylazo-(4-azobenzene)]-pyrrolidine
(S)-MPAAP-N	*(S)*-3-methacryloyloxy-1-[4'-nitrophenylazo-(4-azobenzene)]-pyrrolidine
(S)-MLMA	(S)-maleamic acid-L-leucine methyl ester
(S)-MLSI	(S)-N-succinoyl-L-leucine methyl ester
SIG	Separation or isolation group
SRG	Surface relief gratings
tert-BMA	*Tert*-butyl methacrylate
TGA	Thermogravimetric analysis
TrMA	Triphenylmethyl methacrylate

Abstract

Optically active polymers play a very important role in our modern society. The specialities of optically active polymers are known with their various characteristics as occurred naturally in mimicry. The present review describes the monomers and synthesis of optically active polymers from its helicity, internal compounds nature, dendronization, copolymerization, side chromophoric groups, chiral, metal complex and stereo-specific behaviour. The various properties like nonlinear optical properties of azo-polymers, thermal analysis, chiroptical properties, vapochromic behaviour, absorption and emission properties, thermosensitivity, chiral separation, fabrication and photochromic property are explained in detail. This review is expected to be interesting and useful to the researchers and industry personnel who are actively engaged in research on optically active polymers for versatile applications.

Optically Active Polymers: A Systematic Study on Syntheses and Properties

1 Introduction

Optically active materials are those which can easily rotate a beam of transmitted plane-polarized light into plane of polarization containing unequal amounts of corresponding enantiomers. The origin of optical activity is made in the chiral elements of a polymer such as centres or axes of chiral for long-range conformational order in a macromolecule. In fact, most naturally occurring macromolecules possess the ability to organize to more complex high structure rather than single one and manifest their functions.

The problems of charged and reactive polymers are correlated to optically active compounds and it shows an inherent property of both ordinary macromolecules as well as large range of synthetic polymers. Chiral compounds are optically active and essential for life such as proteins, polysaccharides, nucleic acids, etc. and chirality is most important for existence. About 97 % drugs are formed from natural sources, 2 % are recemates and only 1 % is achiral, in looking of chirality of nearly 800 drugs. Optically active polymers today have also become of great interest and thus play a significant role in molecular assembly and arrangement, which is essential for super molecular structure of optoelectronics [1–4]. The synthetic optically active polymers may also play important role like mimicry of naturally occurring polymers and that's why the extensive studies are required on their synthesis, conformations and properties. Various kinds of optically active polymers e.g., from its helicity, internal compounds nature, dendronization, copolymerization, side chromophoric groups, chiral, metal complex and stereo-specific behaviour are reported, however, those are not placed in a systematic manner. In the present book an effort has been made to collect most of those works in one place for better understanding in the subject with detailed explanation of properties like nonlinear optical properties of azo-polymers, thermal analysis, chiroptical properties, vapochromic behaviour, absorption and emission properties, thermosensitivity, chiral separation, fabrication and photochromism.

© The Author(s) 2017
P.K. Dutta and V. Kumar, *Optically Active Polymers*,
SpringerBriefs in Molecular Science, DOI 10.1007/978-981-10-2606-5_1

– Classification of optically active polymers

Optically active polymers are divided into three types:

a. Biopolymers as obtained from nature.
b. Polymers prepared by almost completely isotactic polymerization by modification of naturally occurring polymer backbones such as polysaccharides.
c. Synthetic polymers as per the requirement with proper tailoring of functional groups.

– Speciality of optically active polymer

The properties of the optically active polymers are similar to other compounds except characteristics chain dimension and structural or conformational changes. Optically active polymers are very important due to its specific properties and attractive applications like creating of complex optical devices, enantiomeric separations as occurred in chromatographic methods as well as dispersion of the specific rotation provides the information regarding Cotton effect or conformational changes of the polymers. Similarly, configurational chirality also offers various kinds of information in terms of asymmetric carbon atom, macromolecular asymmetry and of the asymmetrical centres etc.

– Monomers of optically active polymers

Many biological polymers are collected of different variety but structurally correlated with other monomer residues; e.g. DNA polynucleotide, it contains different subunits of nucleotides. The block copolymers polystyrene—poly(Z-L-lysine) solid-state structures were determined by the polymer architecture and with the help of circular dichroism, electron microscopy, quantitative small and wide-angle X-ray diffraction to confirm the secondary structure of polypeptide [5].

2   Synthesis of Optically Active Polymers

The optically active compounds are synthesized by highly efficient methodologies and catalysts. The various synthetic approaches for optically active polymers are described below:

2.1   Helical Polymer

Helicity is one of the subtlest aspects of polymer chain where the polymer chain spiral structure along the chain axis acts like a spring. Helical polymers are frequently occurring in nature in single, double or triple helices form in genes, proteins, DNA, collagen, enzymes, and polypeptides. The stability of natural polypeptides are increased due to the helical confirmations.

Preparation of artificial helical polymers is a great challenge to the researchers. In this context very little success has been achieved for formation of helical polymers based microscale particles. However, the analogous microparticles as formed from poly(thiophene), poly(phenylene ethynylene), poly(fluorene) and polyacetylenes based conjugated and other vinyl polymers are more pronounced. The work on preparation of nanoparticles from polyacetylenes by Mecking's et al. [6] has drawn very interesting application aspect in inkjet printing. Later on, various groups of researchers have successfully prepared both microparticles and nanoparticles of optically active helical substituted polyacetylenes [6–9]. The optical activity and significant potential applications like asymmetric catalysis, enantiomer-selective crystallization to enantio-selective release, chiral recognition and resolution are remarkable due to the size variations [10–12].

The synthetic helical polymers are to be classified into static or dynamic categories depending on the inversion barrier of the helical conformation [13, 14]. The stability of the static helical polymers are found in solution and energy barriers are relatively high in inversion form, while low energy barriers occur relatively in a mixture of right- and left- handed helical domains which are separated rarely during helix reversals. Like semiconductor impurity, the slight addition of optically active repeating units change the property drastically. The shifting of equilibrium towards excess one-handed helicity occurs with little incorporation of optically active repeating units.

Currently, the properties of biopolymers in terms of chiral recognition with skilled emulating of synthetic helical polymers are the focus of major interest. The applications like catalysis, enantioseparation and sensing are popular responsive three-dimensional intramolecular or intermolecular superchiral structures based molecular recognition. Conjugated optically active polymers are interesting class of chiral macromolecules due to its chiral behaviour and thus those can be utilized for optical or nonlinear electrically conductive properties which are basically for the conjugation along the backbone of the polymer. For example, poly[N-(R)- or (S)-3,7-dimethyloctyl-3,6-carbazole]s (R- or S-PDOC) are the first examples of optically active polycarbazoles, synthesized by using modified nickel coupling method with 60–70 % yield [15–17].

By certain physical factors like thermal, ultraviolet irradiation, high pressure and other chemical parameters like organic solvents the helical polymers are easily denaturalized. A variety of helical polymers are synthesized, which include polyisocyanates, polyisocyanides, polychloral, polymethacrylates, polysilanes, polythiophenes, poly (p-phenylene)s, poly(1-methylpropargyl-ester)s, poly(phenylacetylene)s and poly (-unsaturated ketone) [18–24] (Fig. 1). Other polymers are whose optical activity is main chain or side chain chirality dependent e.g. amino-acid-based polymers are nontoxic, biocompatible and biodegradable.

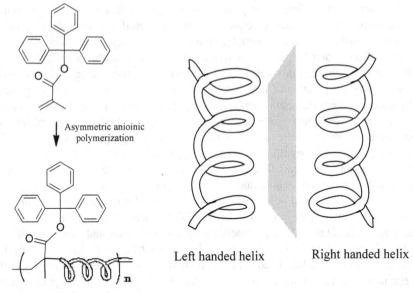

Synthesis of Helical Polymer

Fig. 1 Schematic diagram of synthesis of helical polymers

2.2 Poly (ethyl L-lysinimide)s

The detail synthesis method described elsewhere [13]. In brief, a mixture of dianhydride, ethyl L-lysine dihydrochloride, Et₃N in DMF and their respective concentration 0.001, 0.001, 0.002 mol was stirred at room temperature and then refluxed (Table 1). The product obtained after precipitation, filtration, and drying under vacuum.

2.3 Poly (ethyl L-lysinamide)s

In the same manner as above the reaction mixture of aromatic acid dichloride, ethyl L-lysine dihydrochloride, Et₃N with respective concentration 0.001, 0.001 and 0.004 mol was stirred and the product obtained (Table 1) [25].

2.4 Optically Active Poly(amide–imide)s

The polymer synthesized from 4,4′-(Hexafluoroisopropylidene)-bis(phthaloyl-L-leucine)diacid chloride with 4,4′-(Hexafluoroisopropylidene) N,N′-bis(phthaloyl-L-

Table 1 Structure of monomer, polymer and respective reaction condition

Monomer	Polymer	Reaction condition	Ref. No.
		1. RBr, K_2CO_3, DMF, 50 °C, 24 h; 2. Ni(COD)$_2$, COD, Bpy, DMF, 60 °C, 24 h	[15]
		1. RBr, K_2CO_3, DMF, 50 °C, 24 h; 2. Ni(COD)$_2$, COD, Bpy, DMF, 60 °C, 24 h	[15]
	Poly (ethyl L-lysinmide)	1. Pyromelliticdianhydride, DMF/Et$_3$N, r.t./2 h; reflux/5 h	[15]
		1. Diacylchloride, DMF/Et$_3$N, r.t. 10 h	[25]
		1. Diamine, TPP/Pyridine, $CaCl_2$, NMP	[27]
		1. Aromatic diamines, in microwave oven	[28]

(continued)

Table 1 (continued)

Monomer	Polymer	Reaction condition	Ref. No.
H_3C-CHO		1. KCN, EtOH, Reflux 2. HNO_3, Reflux 3. H_2NCXNH_2, NaOH, EtOH, Reflux 4. $Na_2S_2O_5$, KCN, $(NH_4)_2CO_3$	[30]
TMS—		1. [Rh(nbd)Cl]$_2$, Et$_3$N/anhydrous solvent, 50 °C, 24 h	[30]
		1. R.T., AcOH:Py (3:2) 2. Reflux 3. Δ, $SOCl_2$ 4. o-Cresol/DABCO/microwave irradiation	[33]
		1. 2 NBS/DMF 2. Mg/Et$_2$O 3. FeCl$_3$ in CHCl$_3$	[35]
		1. 2. $H_2C=C-COCl$ 3. AIBN	[55, 56]

(continued)

Table 1 (continued)

Monomer	Polymer	Reaction condition	Ref. No.
		1. 2-Chloroethanol, KOH 2. 4-Nitroaniline + HCL/NaNO$_2$ 3. NaDBS, CH$_2$Cl$_2$/H$_2$O 4. Methacrylyl chloride, THF 5. AIBN, Dichlorobenzene	[67]
(S)-MLMI	(S)-MLSI	1. H$_2$, Pd/C in EtOAc at r.t	[85, 86]
(S)-MLMI	Poly((S)-MLSI)	1. Organometal/chiral ligand 	[85, 86]
R = —CH$_3$ (S)-MLMI R = (S)-BnLMI n-BuLi: Et2Zn: Me2 Zn: —Zn—			
(+)-citronellal		CF$_3$COOH, CHCl$_3$ Zn(OAc)$_2$, DMF	[87]

(continued)

Table 1 (continued)

Monomer	Polymer	Reaction condition	Ref. No.
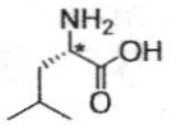	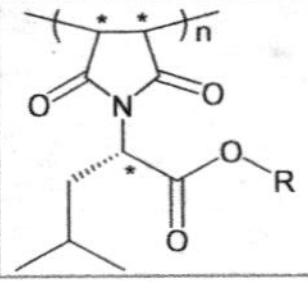	1. Benzyl alcohol, p-TsOH, reflux in benzene 2. Et$_3$N in EtOAc 3. 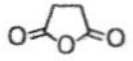 in EtOAc 4. ZnBr$_2$/50 °C, HMDS/80 °C, reflux, 12 h	[89–93]

R = —CH$_3$,

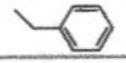

leucine-*p*-amidobenzoic acid) by the condensation polymerization reaction (Table 1). The detail is described elsewhere [26–29]. In brief, after synthesis of monomer 4,4'-(Hexafluoroisopropylidene)-bis(phthaloyl-L-leucine)diacid chloride the optically active poly(amide–imide)s was synthesized by a mixture of 4,4'-(hexafluoroisopropylidene)-*N*,*N*'-bis(phthaloyl-L-leucine-*p*-amidobenzoic acid) (0.1 g), *p*-phenylenediamine (0.012) and CaCl$_2$ (0.014 g) with concentration of 1.10×10^{-4}, 1.10×10^{-4}, 1.26×10^{-4} mol respectively. The mixture was heated at 220 °C under nitrogen atmosphere, poured into methanol and from precipitation the product was collected.

2.5 Poly(phenylacetylene)

4-Iodobenzoyl chloride with phenylalaninol in anhydrous THF with Et$_3$N gives monomer *N*-((*S*)-1-Benzyl-2-hydroxyethyl)-4-iodobenzamide under very mild conditions followed by coupling reaction and thus formed intermediate (*S*)-4-benzyl-2-(4-iodophenyl)oxazoline. It was converted into (*S*)-4-benzyl-2-(ethynylphenyl)-oxazoline (BnEPhOx) using (trimethylsilyl)acetylene (TMSA), in presence of palladium/copper(I) catalyst system in anhydrous THF by Sonogashira–Hagihara coupling reaction (Table 1). The detail preparation method is described elsewhere [30].

2.6 Optically Active Dendronized Polymers

Dendronized polymers are just like consortium and stretched out, anisotropic structure and one molecule of such polymers contains several thousands of dendrons. Depending on dendron generation, the polymers are different.

Hu et al. [31] reported a new type of macromolecular chiral catalysts for asymmetric catalysis using Suzuki coupling polymerization and obtained optically active ephedrine-bearing dendronized polymers. Their finding showed that the optically active dendronized polymers have characteristics joined features like huge numbers of catalytic sites, more solubility and nanoscopic dimensions towards more acceptable in comparison to its existing chiral catalysts of linear polymeric and dendritic nature.

The increasing interest for second-order nonlinear optical (NLO) organic materials are because of their potential applications in electrooptic (E-O) devices especially in high-speed, very broad bandwidth and low driving voltages. Luo et al. [32] synthesized dendronized chromophore containing cardo-type NLO polyimide (PI-CLD) with very high *T*g and very large E-O coefficient high poling efficiency. The prepared material was suitable for E-O device fabrication due to 90 % more efficiency retained at 85 °C for more than 650 h.

2.7   Optically Active Poly(ester-imide)s [33]

When pyromellitic dianhydride (benzene-1,2,4,5-tetracarboxylic dianhydride) [benzo [1,2-c:4,5-c']difuran-1,3,5,7-tetraone] (1) reacted with L-leucine [(S)-2-amino-4-methylpentanoic acid] (2) the resulting amide-acid [4,6-bis(((S)-1-carboxy-3-methylbutyl)carbamoyl)isophthalic acid] (3) undergoes reflux formed imide-acid [N, N'-(pyromellitoyl)-bis-L-leucine diacid] [(2S,2'S)-2,2'-(1,3,5,7-tetraoxopyrrolo[3,4-f] isoindole 2,6(1H,3H,5H,7H)-diyl)bis(4-methylpentanoic acid)] (4). The compound (4) was converted to the N,N'-(pyromellitoyl)-bis-L-leucine diacid chloride [(2S,2'S)-2,2'-(1,3,5,7-tetraoxopyrrolo [3,4-f]isoindole2,6 (1H,3H,5H,7H)-diyl)bis (4-methylpentanoyl chloride)] (5) by reaction with thionyl chloride (Fig. 2). The diacid chloride (5) undergoes polycondensation reaction with several aromatic diols in a facile and rapid reaction process in microwave radiation in the presence of a small amount of a polar organic medium produced good yield and moderate inherent viscosity for optically active poly(ester-imide)s (PEIs). By using benzyl triethyl ammonium chloride as phase transfer catalyst under solution condition the polymerization was completed but it was unsuccessful. Without using the catalyst 1,4-diazabicyclo[2.2.2]octane (DABCO) in the polycondensation reaction of different aromatic diols (6a–6g) with monomer (5), obtained polymers having low inherent viscosities and low yields.

Polycondensation reactions of an equimolar mixture of monomer (5) with different diols (6a–6g) (DABCO used as a catalyst) by microwave assisted method the poly(ester-imide)s, PEIs (7a–7g) were synthesized (Fig. 3).

2.8   Optically Active Polyalkylthiophenes [34–37]

Quinquethiophene monomer: Optically active quinquethiophene monomer contains an enantiomerically pure chiral alkyl group and thiophene ring with a linear C6 alkyl chain at the C-β position, namely, 3,3''''-di n-hexyl-4',3''''-di[(S)-(+)-2-methylbutyl]-2,2':5',2'':5'',2''':5''',2''''-quinquethiophene shown in Fig. 4. The synthesis of β-substituted polythiophenes carried out by β-substituted mono- or oligomeric thiophenic monomers in solution as well as in the solid state by regioselective or regiospecific manner. The polymerization reaction of symmetrically substituted oligothiophenic monomers containing the β-substituents located far apart from the reacting sites whereas asymmetrically substituted monomers occur in highly demanding catalysts and reaction conditions. The polymerization of regioregular macromolecules is more simple and economic with iron (III) trichloride.

By replacement reaction of (S)-(+)-DMBTT, the hydrogen atoms positions at 5 and 5'' of two 3-n-hexyl-2-thienyl moieties form a symmetrically substituted quinquethienyl derivative and produce [(S)-(+)-DHDMBQT], a novel optically active monomer which undergo oxidative mechanism based polymerization of a

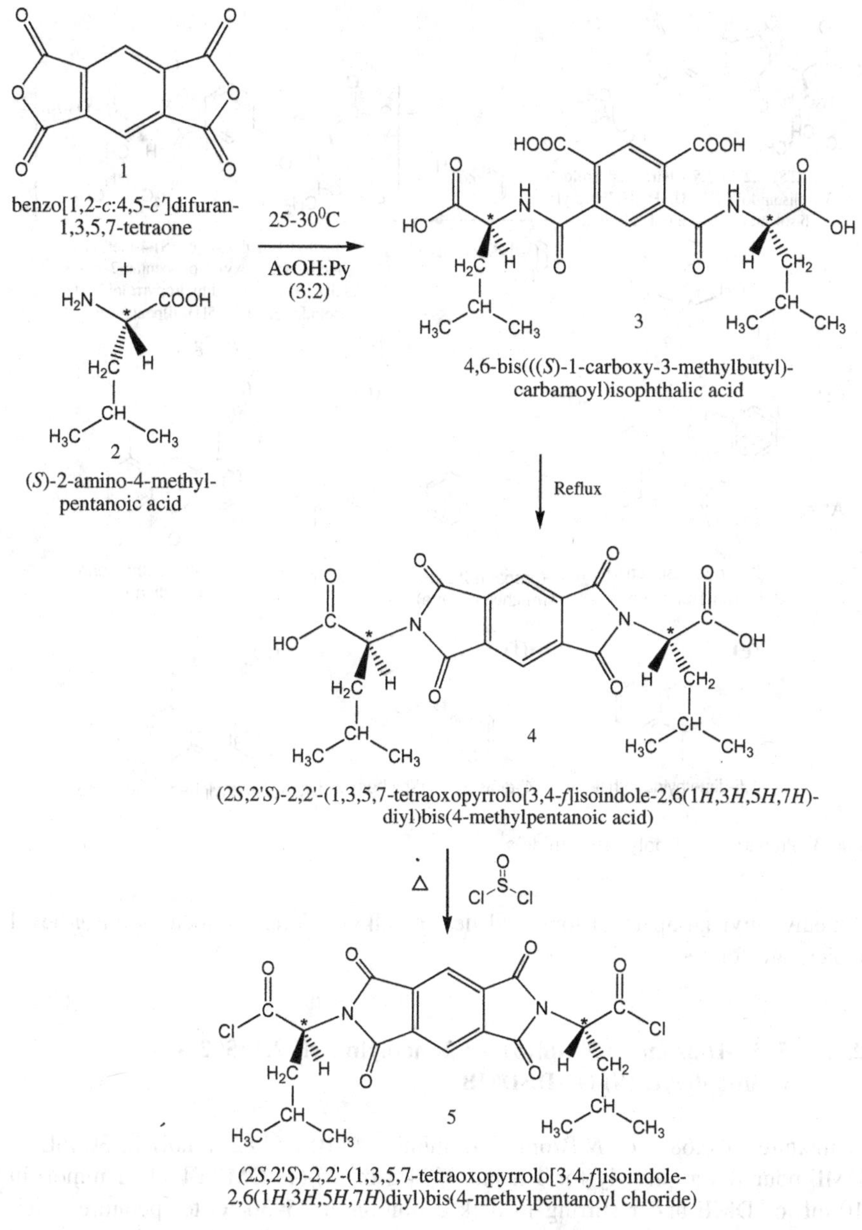

Fig. 2 N,N'-(pyromellitoyl)-bis-L-leucine diacid chloride preparation [33]

regioregular product. While improve the solubility and filmability properties of the poly[(S)-(+)-DHDMBQT] a subsequent polymeric derivative in terms of (S)-(+)-DMBTT of more than 80 % substituted rings and the presence of optically active

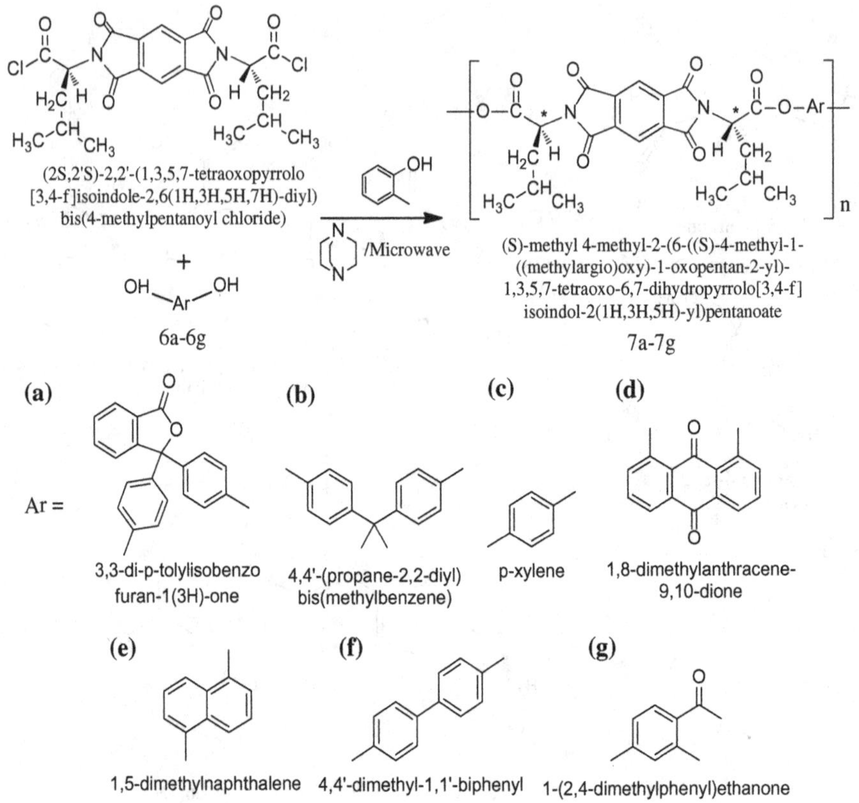

(2S,2'S)-2,2'-(1,3,5,7-tetraoxopyrrolo[3,4-f]isoindole-2,6(1H,3H,5H,7H)-diyl)bis(4-methylpentanoyl chloride)

+

6a-6g

(S)-methyl 4-methyl-2-(6-((S)-4-methyl-1-((methylargio)oxy)-1-oxopentan-2-yl)-1,3,5,7-tetraoxo-6,7-dihydropyrrolo[3,4-f]isoindol-2(1H,3H,5H)-yl)pentanoate

7a-7g

(a)

3,3-di-p-tolylisobenzofuran-1(3H)-one

(b)

4,4'-(propane-2,2-diyl)bis(methylbenzene)

(c)

p-xylene

(d)

1,8-dimethylanthracene-9,10-dione

Ar =

(e)

1,5-dimethylnaphthalene

(f)

4,4'-dimethyl-1,1'-biphenyl

(g)

1-(2,4-dimethylphenyl)ethanone

Fig. 3 Preparation of poly(ester-imide)s

2-methylbutyl group for a long and flexible alkyl side-chain such as the *n*-hexyl moiety are formed.

2.8.1   5, 5″-Dibromo-3,3″-di[(S)-(+)-2-methylbutyl]-2,2′:5′,2″-terthiophene, (S)-(+)-DBDMBTT

A mixture of 0.68 g of *N*-Bromo succinimide (NBS) (3.82 mmol) in 30 mL of DMF poured very slowly for 1 h into 0.74 g of (*S*)-(+)-DMBTT (1.91 mmol) in 10 mL of DMF under stirring in dark condition at 25–30 °C temperature. After 13 h under above condition the reaction mixture was added into cold water and with the help of diethyl ether the reaction mixture was extracted. The solution was washed many times up to removal of ether and then dried by Na_2SO_4. The obtained crude product was evaporated and collected with yield 98 % [34].

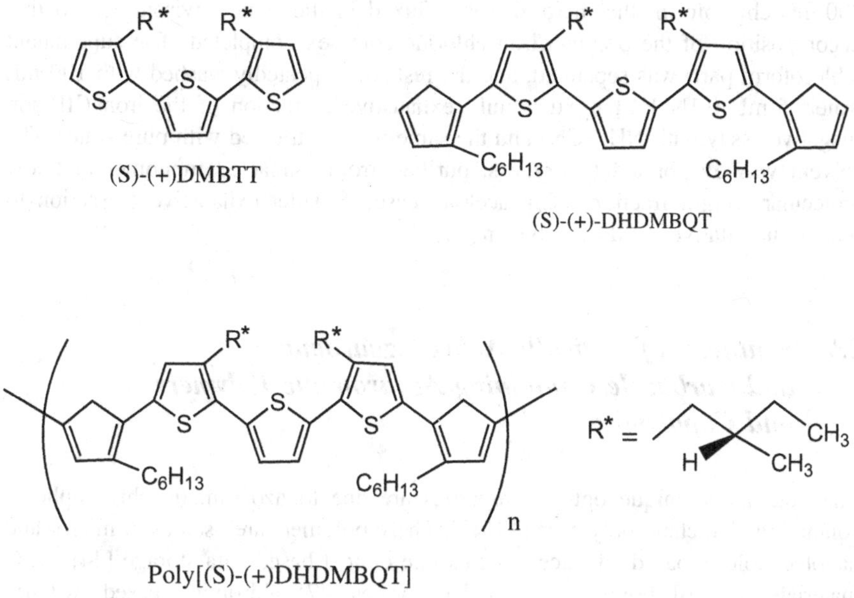

(S)-(+)DMBTT

(S)-(+)-DHDMBQT

Poly[(S)-(+)DHDMBQT]

Fig. 4 Structures of Quinquethiophene monomers and polyalkylthiophenes

2.8.2 3,3''''-Di-N-Hexyl-4',3'''-di[(S)-(+)-2-methylbutyl]-2,2':5',2'':5'',2''': 5''',2''''- quinquethiophene, (S)-(+)-DHDMBQT

In the presence of iodine as a catalyst and under reflux condition for 5 h, the reaction of 2-bromo-3-n-hexylthiophene and magnesium formed a solution of 2-bromomagnesium-3-hexylthiophene in diethylether. After cooling at room temperature, dichloro[1,3-bis(diphenylphosphino)- propane] Ni(II) [NiCl$_2$(dppp)] was added in the reaction and a solution of 5,5''-Dibromo-3,3''-di[(S)-(+)-2-methylbutyl]-2,2': 5',2''-terthiophene in diethyl ether was added drop by drop.

Under reflux condition the reaction mixture was first heated for 20 h, cooled and then poured into 100 mL water containing 2N HCl. With diethylether the synthetic product was repeatedly extracted and the ethereal solution was dried by Na$_2$SO$_4$, finally the crude product was obtained by evaporation under the reduced pressure. Column chromatography (Al$_2$O$_3$, n-hexane) method was used for obtaining pure product. The yield of (S)-(+)-DHDMBQT obtained 0.32 g, 40 % [34].

2.8.3 Polymerization of (S)-(+)-DHDMBQT

A 6×10^{-2} M chloroform solution of one molar (S)-(+)-DHDMBQT and 4 molar anhydrous iron (III) trichloride underwent stirring for 15 h at room temperature with mild flow of nitrogen gas. After the addition of 0.1N HCl (50 mL) and

100 mL chloroform, the mixture was refluxed in the same environment so that decomposition of the polymer/iron chloride complex completed. The supernatant (chloroform part) was separated, and the rest was repeatedly washed with 100 mL water: 5 mL 0.1N HCl mixture until exhaustive extraction of the iron (III) ion (negative assay with NH_4SCN) and the same was neutralized with pure water. The solvent was evaporated to dryness, purified from unreacted monomer and low molecular weight fraction using acetone based Soxhlet exhaustive extraction to obtain quantitative yield of polymer [34].

2.9   Synthesis of Optically Active Azoaromatic and Carbazole Containing Azoaromatic Polymers and Copolymers

The potentially unique optical properties are due to azoaromatic chromophores containing side chain polymers [38–43]. These polymers are used as in micro- and nanotechnology based advance systems like optical based data storage [44], NLO materials [45, 46], holographic based memories [47], chiroptical based switches [48], surface relief based gratings [49–51], etc. The chemical structures with variation of different side chains of the investigated azoaromatic polymers and copolymers are shown in Fig. 5.

Chemical structures of investigated azoaromatic polymers and copolymers, and carbazole containing azoaromatic polymers and copolymers as described by various workers (Angiolini et al., Sperotto et al. and others [52–64]) with detailed methodology and evaluation, a comparative description of their work is elaborated in Table 2.

2.9.1   Homopolymer of (R)-3-methacryloyloxy-1-(4′-cyano-4-azobenzene) pyrrolidine [55, 56]

Synthesis of monomer
(R)-3-methacryloyloxy-1-(4′-cyano-4-azobenzene)pyrrolidine [(R)-MAP-C] was prepared using methacryloyl chloride and (R)-3-hydroxy-1-(4′-cyano-4-azobenzene) pyrrolidine [(R)-HAP-C] (which was derived by the coupling reaction of (R)-3-hydroxy-1-phenyl pyrrolidine [(R)-HPP] and diazonium salt of 4-cyano aniline) in the presence of dimethylamino pyridine (DMAP). The detail method is described elsewhere [65]. In brief, 2.50 g, 8.6 mmol of [(R)-HAP-C] solution in an ice-cooled condition, 0.10 g dimethylamino pyridine as catalyst, and 0.10 g 2,6-di-tert-butyl-4-methyl phenol as an inhibitor, 52 mL dry methylene dichloride and 1.00 mL, 10.3 mmol methacryloyl chloride in 5 mL methylene dichloride and 1.44 mL, 10.3 mmol triethylamine in 5 mL methylene dichloride were simultaneously added under nitrogen environment. The reaction mixture was

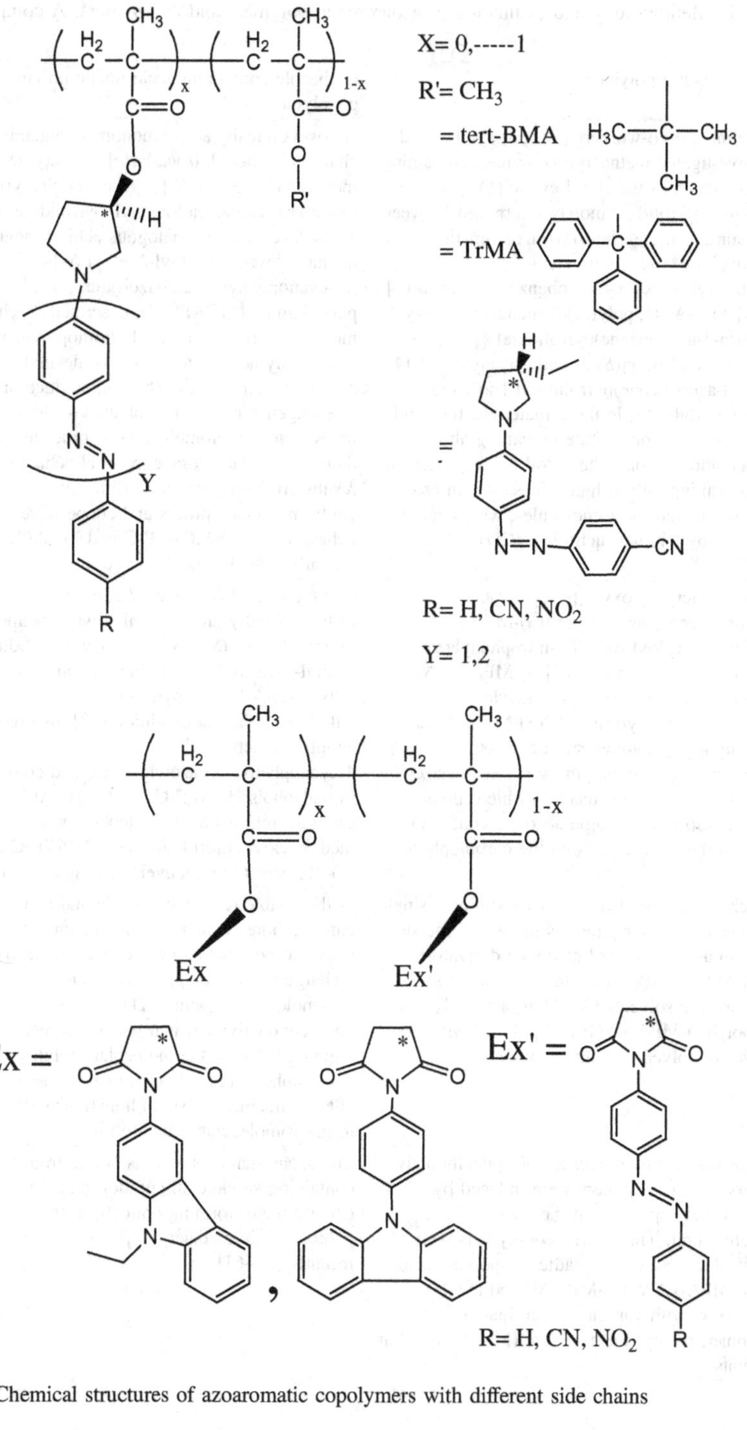

Fig. 5 Chemical structures of azoaromatic copolymers with different side chains

Table 2 Methodology and evaluation of azoaromatic polymers and copolymers: A comparative study

Bisazoaromatic polymers	Carbazole containing azoaromatic polymers and copolymers
Angiolini and co-workers [52, 53] synthesized and investigated methacrylic polymers containing the side chain in the chiral cyclic (S)-3-hydroxypyrrolidine moiety interposed between the main chain and the *trans* azoaromatic chromophore, such as poly[(S)-3-methacryloyloxy-1-(4-azobenzene)pyrrolidine] {poly[(S)-MAP]}, poly[(S)-3-methacryloyloxy-1-(4'-nitro-4-azobenzene)pyrrolidine] {poly[(S)-MAP-N]} and poly[(S)-3-methacryloyloxy-1-(4'-cyano-4-azobenzene)pyrrolidine] {poly[(S)-MAP-C]} [54–56]. In these materials, the rigid chiral moiety is one of the prevailing absolute configuration favours the chiral conformation of one prevailing helical handedness within chain segments of the macromolecules, which can be observed by circular dichroism (CD)	A novel optically active monomer containing side chain in 9-phenyl-azocarbazole moiety, (S)-methacryloyl-3-oxy-N-{4-[(4 cyanophenyl)-(3-carbazoil)-diazene]-phenyl}pyrrolidine ((S)-MCAPP-C) and its analogous achiral monomer methacryloyl-2-oxyethyl-N-ethyl-N-{4-[(4-cyanophenyl)(3-carbazoil)diazene]}-phenylamine (MCAPE-C), described. Each monomer has been radically homopolymerized and copolymerized to afford the desired polymeric derivatives. The photoinduction of birefringence has been evaluated on films of the investigated macromolecules in order to assess their optical data storage material behaviour. The results are interpreted with different conformational stiffness and cooperative behaviour of (S)-MCAPP-C and MCAPE-C azocarbazole chromophoric co-units
(S)-3-methacryloyloxy-1-[4'-cyanophenylazo-(4-azobenzene)]-pyrrolidine [(S)-MPAAP-C], (S)-3-methacryloyloxy-1-[4'-nitrophenylazo-(4-azobenzene)]-pyrrolidine [(S)-MPAAP-N] and (S)-3-methacryloyloxy-1-[4'-phenylazo-(4-azobenzene)]-pyrrolidine [(S)-MPAAP] are optically active monomers, it and its corresponding homopolymers was synthesized by radical polymerization reaction which contained a chiral group in one single absolute configuration between the main chain and the chromophore	The mixture of N-bromo-N-ethyl-N-(2-hydroxyethyl)amine (which was obtained by bromuration of (S)-(+)-3-hydroxy-pyrrolidine and Nethyl-N-(2-hydroxyethyl)amine) and (S)-(+)-3-hydroxy-N-bromo-pyrrolidine, under Ullmann's reaction condition [57] undergoes to coupling reaction with 4-cyanophenylcarbazolyldiazene and converted into alcohols HCAPE-C and (S)-HCAPPC. After esterification of obtained alcohols with methacryloyl chloride formed MCAPE-C and (S)-MCAPP-C respectively with good yields
The characterization and properties of only single azoaromatic chromophore as present in the side chain in above reported analogue derivatives: Poly[(S)-MPAAP] was soluble in commonly used organic solvents like THF or CHCl₃, while the poly[(S)-MPAAP-C] and poly[(S)-MPAAP-N] was dissolved in limited solvents	In the visible region, carbazole and azo chromophores containing side chain optically active co-polymers give rise to charge carried via a charge transfer complex of induced intramolecular structure. Due to the photoconductivity which was concerned to hole hopping between the molecular mobility of chromophore and side chain carbazole units affected the mechanism of hole transportation due to macromolecular structure [58]
The photochromic properties of optically active bisazoaromatic polymers were induced by progressively spacing out the backbones of repeating unit. The desired copolymers were obtained respectively by radical copolymerization of (S)-MPAAP-N, (S)-MPAAP and (S)-MPAAP-C with variation of an inactive *co* monomer, methyl methacrylate (MMA) in molar amounts	The azoaromatic as well as chiral groups containing single configuration linked to the side chain directly forming optically active photochromic methacrylic polymers are reported recently [59–61]

(continued)

Table 2 (continued)

Bisazoaromatic polymers	Carbazole containing azoaromatic polymers and copolymers
The optical data storage is evaluated by photo induction of birefringence on thin films of the investigated copolymers. The bisazoaromatic chromophoric *co*-units are interpreted as conformational stiffness and composition of copolymers, due to this it shows poor solubility and slower optical response rates comparatively having only single azo bond in analogous polymers. These polymers having properties of larger thermal and temporal stability, due to this it was using in nanoscale technology for potential use in optoelectronics and in the manipulation of all-optical data	Dissymmetric systems shows exciton splitting of dichroic absorptions and optical activity and photochromic materials having NLO, photoresponsiveness and photorefractivity properties, which occurred due to the presence of azoaromatic and chiral functionalities in the polymers [62]. Finally, novel optically active multifunctional methacrylic copolymers were synthesized, which contained side-chains chiral moieties and linked to a photoconductive carbazolic and to a photochromic azoaromatic chromophores
The effect on chiroptical and thermal properties due to the presence of achiral comonomers into polymer and its characterization in terms of side-chain mobility and grown hindrance was studied. The monomer (*S*)-MPAAP with triphenylmethyl methacrylate (TrMA) and an inactive *tert*-butyl methacrylate (*tert*-BMA) in different concentrations followed by radical copolymerization reaction. At different concentration of (*S*)-MPAAP *co*-units, the structures of the synthesized compounds poly [(*S*)-MPAAP-*co*- TrMA]s and poly[(*S*)-MPAAP-*co-tert*-BMA]s were investigated	The monomer *trans*-(*S*)-(+)-2-methacryloyloxy-*N* (4-azobenzene)-succinimide (*S*)-(+)-MOSI with (*S*)-(+)-methacryloyl-2-oxy-*N*-[3-(9-ethylcarbazole)]-succinimide (S)-(+)-MECSI and the same monomer with (S)-(+)-methacryloyl-2-oxy-*N*-9-phenylcarbazole-succinimide (*S*)-(+)-MCPS followed radical copolymerization reaction. Further copolymerization of the monomers (*S*)-3-methacryloyloxy-1-(4′-nitro-4-azobenzene) pyrrolidine (*S*)-MAP-*N* and (*S*)-3-methacryloyloxy-1-(4′-ciano-4-azobenzene) pyrrolidine (*S*)-MAP-C with (*S*)-(−)-methacryloyl-3-oxy-*N*-[3-(9-ethylcarbazole)]-pyrrolidine (*S*)-(−)-MECP and (*S*)-(−)-methacryloyl-3-oxy-*N*-9-phenylcarbazole-pyrrolidine (*S*)-(−)-MCPP respectively were studied [63, 64]. The spectroscopic and thermal properties of those copolymeric products were compared to respective monomers and homopolymers
First (*R*)-3-hydroxy-1-phenyl pyrrolidine [(*R*)-HPP] reacted with 4-cyano aniline via coupling to form diazonium salt, (*R*)-3-hydroxy-1-(4′-cyano-4-azobenzene) pyrrolidine [(*R*)-HAP-C]. The salt as an intermediate was allowed to react with methacryloyl chloride to synthesize novel optically active monomer (*R*)-MAP-C as shown in Table 1	The optically active azocarbazolic chemical multifunctional polymers may find multipurpose applications for commercial NLO based devices, photo refractivity and other optical based information systems

ice-cooled for 2 h, left for one night in ambient temperature, and finally washed in order of HCl, Na_2CO_3 and with water. The non-aqueous layer of the reaction mixture was dried and finally the solvent was vaporized. The product (*R*)-MAP-C was purified, yield 87 % and melting point 123 to 125 °C [65].

AIBN: monomer (2 % w/w) was used as free radical thermal initiator in 15 mL of THF solutions reaction vials. Under nitrogen atmosphere the polar composition was prepared and injected into the vials, and freeze-thaw cycles were followed to eliminate any trace amount of dissolved oxygen under 60 °C for 72 h. The excess amount of methanol (100 mL) was poured into the reaction mixture, filtered and the product obtained. The product again dissolved in DMF and, methanol was used to re-precipitate the product. The unreacted monomers were removed by extraction technique. At 70 °C under vacuum condition for several days the product was dried until the constant weight achieved. The obtained dried polymers were highly soluble in polar solvents. Gravimetric method was used for the determination of conversions.

2.9.2 Optically Active Methacrylic Polymer and Azocarbazole Chromophore as Side-Chain: Poly[(S)-MLECA]

(S)-(4-Cyanophenyl)-[3-[9-[2-(2-methacryloyloxy-propanoyloxyethyl] carbazolyl]] diazene [(S)-MLECA] chiral monomer containing the carbazole moiety is a novel optically active photochromic polymer. After study of chiroptical, thermally and spectroscopic properties shows the presence of dipolar interaction characteristics between chromophores and chiral confirmation of prevailing helical hardness for chain segments of macromolecules.

In the presence of sodium dodecyl benzenesulfonate the coupling of commercial 9-(2-hydroxyethyl)carbazole with 4-cyanobenzenediazonium chloride in biphasic medium (H_2O–CH_2Cl_2) give (4-cyano phenyl)-[3-[9-(2-hydroxyethyl)carbazolyl]] diazene [HECA] [66].

(S)-MLECA was prepared by the esterification reaction of HECA and (S)-(−)-methacryloyl-L-lactic acid with a coupling agent 4-(diphenylamino)piridinium 4-toluensulfonate (DPTS) and condensation activator N,N-diisopropyl-carbodiimide (DIPC), the high molecular weight polyesters preparations method described elsewhere [67, 68]. The yield of product was 19 %. The above radical homopolymerization using AIBN as thermal initiator and similar process was followed to obtain the polymer.

2.9.3 Methacrylic Polymers with 9-Phenylazocarbazole Moieties as Side Chains

(S)-3-Hidroxy-N-{4-[(4-cyanophenyl)-(3-carbazolyl)-diazene]-phenyl}-pyrrolidine [(S)-HCAPP-C]

The detail preparation method is described elsewhere [57]. In brief, 4-cyanophenyl-3-carbazolyl-diazene, trans-cyclohexanediamine, trans-cyclohexanediamine and air-stable CuI (in dioxane) were mixed in presence of K_2CO_3, then (S)-(−)-3-hydroxyethyl-N-(4-bromophenyl) pyrrolidine in dioxane was added and refluxed for 24 h under inert atmospheric condition. The product, [(S)-HCAPP-C] was obtained after cooling, filtration and precipitation in MeOH with yield of 80.0 %.

2-Hydroxy-N-ethyl-N-{4-[(4-cyanophenyl)-(3-carbazolyl)-diazene]} phenyl-amine [HCAPE-C]

The reaction of 2-hydroxyethyl-N-ethyl-N-(4-bromophenyl)amine and 4-cyanophenyl-3-carbazolyl-diazene produced [(S)-HCAPP-C], the detailed procedure described elsewhere [57]. The yield of pure HCAPE-C was 69 %.

By esterification of the related alcohols with methacryloyl chloride the (S)-MCAPP-C and MCAPE-C respectively were obtained with good yields [57]. The polymerization of monomers was followed by above AIBN thermal initiation to obtain the polymer.

2.9.4 Optically Active Copolymers with Carbazole Side Chain and Azochromophores

In the main chain of the polymer at the 3-position was linked into the side chain carbazole and thus it can produce charge through an induced intramolecular charge transfer complex in the visible region. The hole transport mechanism in macromolecular structure of these materials was affected because of photoconductivity and hole hopping related between the molecular mobility of chromophore and the side chain carbazole units, and thus, the chemical anchorage to the polymer backbone generally depressed [58].

In recent studies reported the optically active photochromic methacrylic polymers bearing two distinct functional groups (i.e. azoaromatic and a chiral group of one single configuration) directly linked to the side chain [59]. In these materials, the presence of a rigid chiral moiety of one prevailing absolute configuration favours the establishment of a chiral conformation of one prevailing helical handedness, at least within chain segments of the macromolecules, which can be observed by circular dichroism (CD) [60].

Azoaromatic polymers shows many characteristics e.g. excitation splitting of dichoric absorption, optical activity, photoresponsiveness, photorefractivity and NLO properties [62].

Henceforth, the synthesis of multifunctional methacrylic copolymers with side-chains chiral moieties linked to a photochromic azoaromatic and to photoconductive carbazolic chromophores based novel optically active polymers are very meaningful.

The radical copolymerization of the respective monomers, (S)-(+)-methacryloyl-2-oxy-N-[3-(9-ethylcarbazole)]-succinimide, (S)-(+)-MECSI and (S)-(+)-methacryloyl-2-oxy-N-9-phenylcarbazole-succinimide, (S)-(+)-MCPS were made with trans-(S)-(+)-2-methacryloyloxy-N(4-azobenzene)-succinimide, (S)-(+)-MOSI and copolymerization of the monomers, (S)-3-methacryloyloxy-1-(4'-nitro-4-

azobenzene) pyrrolidine, (S)-MAP-N and (S)-3-methacryloyloxy-1-(4'-ciano-4-azobenzene)pyrrolidine, (S)-MAP-C with (S)-(−)-methacryloyl-3-oxy-N-[3-(9-ethylcarbazole)]-pyrrolidine, (S)-(−)-MECP and (S)-(−)-methacryloyl-3-oxy-N-9-phenylcarbazole-pyrrolidine (S)-(−)-MCPP respectively were reported. The characterization of copolymeric products were done by thermal and spectroscopic method and compared with respective optically active monomers and homopolymers [63, 64].

This new special class of optically active azocarbazolic multifunctional polymers bearing several chemical functionalities at the same time having properties of all optical manipulation of information, photo refractivity and a variety of potential commercial applications as such as NLO are known today.

Optically Active Side Chain 3-Carbazole Based Copolymers and Azochromophores: Poly[(S)-(+)-MECSI-Co-(S)-(+)-MOSI]s and Poly[(S)-(−)-MECP-Co-(S)-MAP-N]

Azoaromatic and 3-carbazole groups functionalized methacrylic copolymers linked to a chiral group of 2-hydroxysuccinnimide and 3-hydroxypyrrolidine and thus two novel optically active polymers with prevailing absolute configuration have been prepared and characterized.

The glass transition temperature found around 200 °C and high decomposition temperatures. The presence of dipolar interactions between side chain moieties and occurrence of confirmation of chirality in chain segments of macromolecules is characterized by chiroptical, thermal and spectroscopic methods.

Synthesis and structural characterization of copolymers: The desired polymers were obtained by copolymerization of (S)-(+)-MECSI and (S)-(−)-MECP with [(S)-(+)-MOSI] and [(S)-MAP-N] respectively in THF solution using AIBN initiator based radical polymerization. The obtained copolymers were compared with earlier reported poly[(S)-(+)-MECSI], poly[(S)-(−)-MECP] [68], poly[(S)-(+)-MOSI] and poly[(S)-MAP-N] with their relevant data.

The formation of copolymers with respect to the corresponding monomers were determined by IR spectroscopic technique where disappearance of the methacrylic double bond (around 1630–1636 cm^{-1}), and the contemporary shift (10–20 cm^{-1}) of the estereal carbonyl stretching to higher frequencies (1725–1732 cm^{-1}) were shown due to the reduced electron delocalization.

The liquid crystalline behaviour of a thin film of the copolymers was not observed by polarising microscope whereas second-order transitions originated glass transition temperature (T_g), with no melting peaks, were observed in DSC thermograms of the investigated polymers. The T_g values in the range 202–205 °C appear quite high can be attributed to high thermal stability, and the materials can be of interesting for optoelectronics applications. Succinimide and pyrrolidine ring makes the polymer backbone conformationally rigid as present between the chromophores [69, 70], as well as the higher T_g also shows the rigid structure and thus reduces the mobility of the macromolecular chains.

Optically Active Side Chain 9-Phenylcarbazole Based Copolymers and Azochromophores: poly[(S)-(+)-MCPS-Co-(S)-(+)-MOSI], poly[(S)-(−)-MCPP-Co-(S)-MAP-N] and poly[(S)-(−)-MCPP-Co-(S)-MAP-C]

Optically active bi-functionals co-polymers poly[(S)-(−)-MCPP-co-(S)-MAP-C], poly[(S)-(−)-MCPP-co-(S)-MAP-N] and poly[(S)-(+)-MCPS-co-(S)-(+)-MOSI] were synthesized and their structural characterization were made. The application of this synthesized polymers are found in chiroptical switches, photoconductive as well as photorefractive and photonic materials for NLO and optical storage.

The uses of optically active co-monomers: (S)-(+)-MCPS and (S)-(−)-MCPP as holes transport side-chain and 9-phenylcarbazole derivatives (S)-(+)-MOSI, (S)-MAP-N and (S)-MAP-C as electro-optic side-chain known today. The dipolar interactions between chromophores and an ordered chiral confirmation of prevailing helical hardness are characterized by thermally and spectroscopic method.

Synthesis and structural characterization of copolymers

Optically active monomers (S)-(+)-MECPS and (S)-(−)-MCPP were copolymerized in THF solution (using AIBN as an initiator) with (S)-(+)-MOSI and (S)-MAP-N respectively. Then obtained copolymerized products were dissolved in THF solution after using methanol for reprecipitation, finally exhaustive Soxhlet extraction with methanol followed by acetone method used for purification from monomeric and oligomeric impurities. FTIR, ^{1}H- and ^{13}C-NMR were used for the confirmation of structures of synthesized products [14].

2.10 Synthesis of Optically Active Benzodithiophene Based Poly(aryleneethynylene)s for Solar Cells [71]

Ethynylene-substituted benzo [1,2-b:4,5-b']dithiophene (BDT) was synthesized and characterized and its application construction of poly(aryleneethynylene)s (PAEs), alkoxy-substituted BDT and thiazolothiazole were selected as the other copolymerized units for building blocks. Polymers, PEBBDT and PEBTTZ used for applications to organic solar cells and good optical devices were synthesized by Pd-catalyzed Sonogashira coupling reaction. Easily synthesized and chemically stable conjugated poly(aryleneethynylene)s (PAEs) are an interesting class of polymers due to its interchain interactions into rigid backbones. The monomers are synthesized by following routes/steps as given below:

2.10.1 Ethyl 2-amino-5-octylthiophene-3-carboxylate

The detail preparation method is described elsewhere [71]. In brief, Decanal, ethyl cyanoacetate and sulphur dissolved in EtOH, with the addition of triethylamine as base, the reaction occurred at 60 °C for a day. The product of 100 % as yellowish colour oil obtained after purification and removal of solvent.

2.10.2   Ethyl 5-Octylthiophene-3-Carboxylate

The reaction mixture of ethyl 2-amino-5-octylthiophene-3-carboxylate in suspension solution of t-BuONO and $CuCl_2$ in EtOH was stirred for 0.5 h and then ammonium chloride solution added and stirred for 15 min. After removal of solvent, residue partitioned between EtOAc and water, drying by Na_2SO_4 and purification the product obtained as yellow oil with 70 % yield [71].

2.10.3   5-Octylthiophene-3-carboxylic acid

A mixture of ethyl 2-amino-5-octylthiophene-3-carboxylate and NaOH in EtOH was refluxed for 3 h. After removal of solvent, water was added in the residue and pH adjusted to 1 by HCl. After filtration, the obtained precipitate was recrystallized from hexane. Final compound was 5-octylthiophene-3-carboxylic acid was a white solid and yield was 80 % [71].

2.10.4   *N,N*-Diethyl-5-octylthiophene-3-carboxamide

The reaction mixture of 5-octylthiophene-3-carboxylic acid, methylene chloride, and oxalyl chloride were cooled and then brought to ambient temperature with stirring for a day, a clear solution obtained. The obtained thiophene-3-carbonyl chloride, diethylamine and methylene chloride were mixed in ice-water bath, and the reaction was stirred at ambient temperature, washed several times with water and anhydrous Na_2SO_4 was used into the organic layer. The obtained product was pale yellow oil, purified by column chromatography after removing the crude product and yield was 90 % [71].

2.10.5   2,6-Dioctyl-4,8-dihydrobenzo[1,2-b:4,5-b′]dithiophen-4,8-dione

The detail preparation method is described elsewhere [71]. In brief, *N,N*-diethyl-5-octylthiophene-3-carboxamide dissolved in THF under an inert atmosphere and the solution was cooled down by an ice-water bath. *n*-Butyllithium was added into the flask dropwise with stirring at ambient temperature for 30 min. The entire reactant was poured into ice water and stirred for several hours. Finally obtained product was filtrated, successively washed by using water, methanol and hexane. The yield of obtained product as a yellow solid was 72 %.

2.10.6 2, 6-Dioctyl-4, 8-bis (trimethylsilylethynyl)-benzo[1, 2-b:4, 5-b'] dithiophene

The detail preparation method is described elsewhere [71]. In brief, before adding quinone solution, the reaction mixture of *n*-butyllithium and acetylene in THF solution was kept at 0 °C under argon atmosphere and stirred for a day until room temperature. The reaction mixture was quenched by HCl and further stanous chloride dehydrate in acetic acid was added to obtained the product. The yield of obtained product 2, 6-dioctyl-4,8-bis(trimethylsilylethynyl)-benzo[1,2-b:4,5-b'] dithiophene as a white colour solid was 32 %.

2.10.7 2, 6-Dioctyl-4,8-diethynyl-benzo[1,2-b:4,5-b']dithiophene

The detail preparation method is described elsewhere [71]. In brief, 6-dioctyl-4,8-bis(trimethylsilylethynyl)-benzo[1,2-b:4,5-b']dithiophene and Bu_4NF in THF solution was quenched by water and extracted. After several treatment processes, the product was obtained as a white colour solid with 81 % yield.

The chemical structures of investigated monomers as described above are shown in Fig. 6.

2.10.8 Synthesis of Polymer PEBBDT

The detail preparation method is described elsewhere [71]. In brief, 2,6-dioctyl-4,8-diethynyl-benzo[1,2-b:4,5-b']dithiophene, 4,8-dialkoxy benzo [1,2-b:4,5-b']-2,6-di-bromodithiophene, $Pd(PPh_3)_4$ and CuI refluxed with argon gas. Toluene and diisopropylamine added at room temperature in the reaction mixture, stirred for a day at high temperature. The product PEBBDT was obtained after precipitation in MeOH and centrifugation with yield of 74.0 %, GPC: $M_n = 18.5$ K; $M_w/M_n = 1.5$.

2.10.9 Synthesis of Polymer PEBTTZ

Similarly, PEBTTZ was synthesized by following the same as above except 2,5-bis (3'-bromo-5'-alkyl-thiophen-2'-yl)thiazolo[5,4-d]thiazole instead of 4,8-dialkoxy benzo [1,2-b:4,5-b']-2,6-di-bromodithiophene. The product obtained as solid red product with 26 % yield [71].

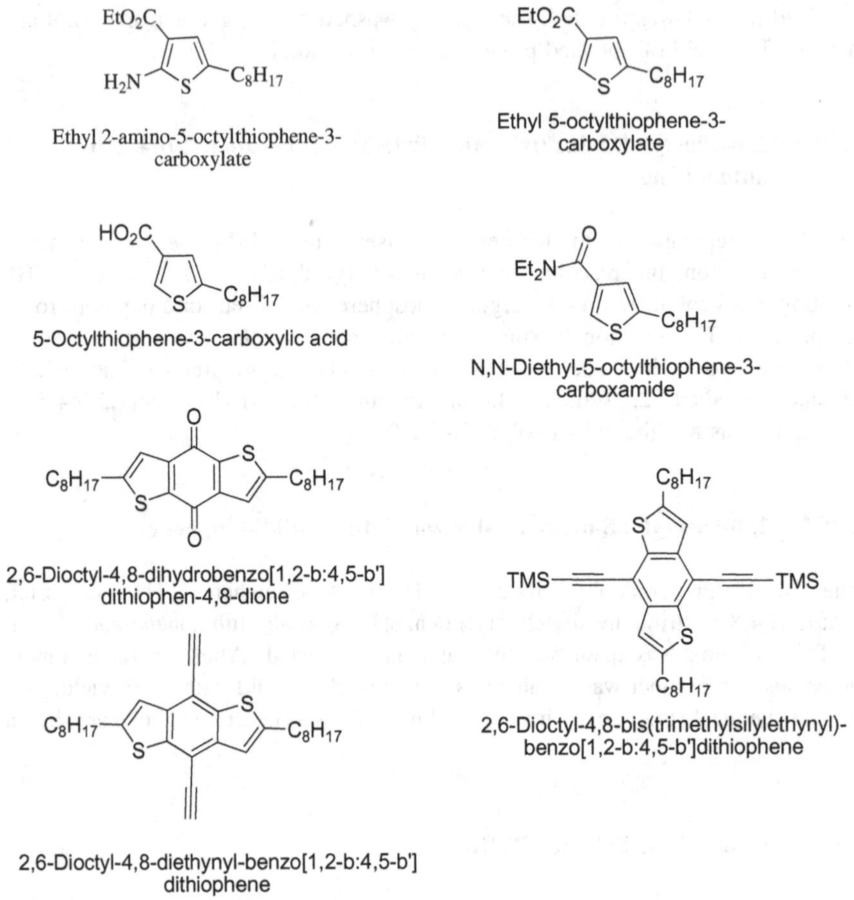

EtO$_2$C
H$_2$N   S   C$_8$H$_{17}$

Ethyl 2-amino-5-octylthiophene-3-
carboxylate

EtO$_2$C
S   C$_8$H$_{17}$

Ethyl 5-octylthiophene-3-
carboxylate

HO$_2$C
S   C$_8$H$_{17}$

5-Octylthiophene-3-carboxylic acid

Et$_2$N   O
S   C$_8$H$_{17}$

N,N-Diethyl-5-octylthiophene-3-
carboxamide

C$_8$H$_{17}$ S   O   S   C$_8$H$_{17}$
O

2,6-Dioctyl-4,8-dihydrobenzo[1,2-b:4,5-b']
dithiophen-4,8-dione

C$_8$H$_{17}$
S
TMS   TMS
S
C$_8$H$_{17}$

2,6-Dioctyl-4,8-bis(trimethylsilylethynyl)-
benzo[1,2-b:4,5-b']dithiophene

C$_8$H$_{17}$ S   S   C$_8$H$_{17}$

2,6-Dioctyl-4,8-diethynyl-benzo[1,2-b:4,5-b']
dithiophene

Fig. 6 Chemical structure of investigated monomers

2.10.10 Synthesis and Structural Characterization

The reaction of 2,6-dioctyl-4,8-diethynyl-benzo[1,2-b:4,5-b']dithiophene with dibromide compounds in toluene by Pd/Cu-catalyzed Sonogashira coupling copolymerization produced the polymers PEBBDT and PEBTTZ. Thiophene-3-carboxylic acid was similarity converted into the 2,6-Dioctyl-4,8-diethynyl-benzo [1,2-b:4,5-b']dithiophene as the synthesis of common BDT compound [72]. Following a three-component reaction, aldehyde, sulfur, and cyanoacetate of the 2-alkyl substituted thiophene unit was converted into 2-amino thiophene [73]. The compound was obtained in good yield by the reaction of deamination and hydrolysis. PEBBDT was readily soluble in organic solvents like chloroform, THF, chlorobenzene (CB), etc. at room temperature but PEBTTZ was less soluble in these solvents [71].

2.11   Synthesis of Optically Active Porphyrin Derivatives Bearing Four Chiral Citronellal Moieties

Optically active porphyrin derivatives have drawn the attention of biologists relevance and industrialist due to its wide range of applications in the fields of nonlinear optics and chiral catalysis [74–80]. Subsequently, a number of chiral amino acids and chiral hydrocarbons based porphyrin-chiral substituents molecules have been synthesized. It is interesting to note that the presence of stereocenters onto the periphery of porphyrin molecule provides just a possibility for fabrication of helical supramolecular structures and optically active porphyrin molecules.

In contrast, many peripheral substituents of porphyrin compounds with stereocenters even do not display any CD signals. The whole porphyrin absorption range at the molecular level i.e., from the peripheral chiral substituents with stereocenters to the porphyrin chromophore possess weak perturbation [81, 82].

Similarly, many porphyrin derivatives could only form normal supramolecular structures without exhibiting helicity [83, 84]. As a result the increase of chiral perturbation from the peripheral substituents with stereoenters to the porphyrin chromophore forms artificial optically active porphyrin derivatives. Tamura et al. [85] reported preparation of porphyrin derivative with four chiral sugar chains. Reversely, the porphyrin Soret band region on aggregate due to synergistic interplay of intermolecular peripheral chiral porphyrin substituent (pep) undergoes interaction between sugar chains of porphyrin rings and hydrogen bonding and thus shows a perfect Cotton effect. Very recently, the formation of elemental one-dimensional helices from an optically active porphyrin compound bearing four peripheral chiral binaphthyl moieties, linked at the meso-phenyl substituents through crown ether moieties have been reported by Lu et al. [86]. The synthesis of optically active porphyrin derivatives and forming those into helical supramolecular structures still remains unclear.

2.11.1   Metal Free Meso-5,10,15,20-tetra-(2,6-dimethyl-5-heptenyl)porphyrin

The detail preparation method is described elsewhere [79]. In brief, pyrrole and citronellal were mixed in chloroform and trifluoroacetic acid was added into the reaction mixture and kept in dark for a day then DDQ added. The product meso-5,10,15,20-tetra-(2,6-dimethyl-5-heptenyl)porphyrin was obtained by evaporation under reduced pressure and purification as a red-brown viscous liquid with yield of 10.0 %.

2.11.2   Meso-5,10,15,20-Tetra-(2,6-dimethyl-5-heptenyl) porphyrinato Zinc Complex

Similarly, $Zn(OAc)_2.2H_2O$ and   free   meso-5,10,15,20-tetra-(2,6-dimethyl-5-heptenyl)porphyrin in DMF mixed and refluxed under nitrogen. The product meso-5,10,15,20-tetra-(2,6-dimethyl-5-heptenyl) porphyrinato zinc complex was obtained by removal of solvent in vacuo and purification as dark green in colour with yield of 85.0 % [79]. Lu et al. [87] was synthesize a new optically active metal free porphyrin (R)-enantiomer linked into four chiral citronellal units at the meso positions on the porphyrin ring and its zinc congener.

The effective chiral information transfer from the chiral citronellal tails to porphyrin chromophore at the molecular level was compared with the CD silent of N,N',N''-Tris{21H,23H-5-p-aminophenyl-10,15,20-tris-[p-(S)-3,7-dimethyloctoxyp henyl] porphyrin}-1,3,5-tricarboxamid, possess similar chiral units linked onto the meso positions of the porphyrin ring via meso-attached benzene and both compounds show a positive CD signal in the Soret absorption region of both porphyrin compounds. The study reveals the new insight into chiral information transfer as well as the effect of substituents position on the asymmetrical perturbation and expression for synthetic conjugated systems at molecular and supramolecular level.

2.12   Asymmetric Polymerizations of N-Substituted Maleimides Bearing L-Leucine Ester Derivatives

Asymmetric polymerizations of achiral or chiral N-substituted maleimides (RMIs) to obtain optically active polymers and optical resolution abilities of the poly (RMI) s were systematically investigated by Oishi et al. [88]. (S)-N-maleoyl-L-leucine methyl ester ((S)-MLMI) and (S)-N-maleoyl-L-leucine benzyl ester ((S)-BnLMI) both N-substituted maleimides bearing L-leucine ester derivatives ((S)-RLMI) were synthesized from maleic anhydride, L-leucine, and corresponding alcohols. N-substituents, initiators, solvents and temperature influenced the specific rotations and chiroptical properties and structures of poly((S)-RLMI)s were evaluated by GPC, CD, and NMR techniques. Excessive chiral centers of the main chain was induced through anionic polymerizations in addition to the chirality of the N-substituents to achieve optical activities of poly((S)-RLMI)s [88–92]. An unsaturated reactive group in the side ester group of RMI of (S)-N-maleoyl-L-leucine propargyl ester and (S)-N-maleoyl-L-leucine allyl ester both undergo asymmetric polymerization. The optical resolutions of chemically bonded-type chiral stationary phases (CSPs) of those polymers were discussed [93, 94].

The researchers choose L-leucine as chiral amino acid and introduce methyl group and benzyl group respectively to the carboxyl group of L-leucine to extend their research area. The syntheses and asymmetric polymerizations of (S)-N-aleoyl-L-leucine methyl ester, ((S)-MLMI) and (S)-N-maleoyl-L-leucine benzyl ester, ((S)-BnLMI) and other two kinds of RMIs bearing L-leucine ester derivatives, and their applications as CSPs for HPLC are described. The thorough evaluation of chiroptical properties and structures of these polymers were made by using nuclear magnetic resonance (NMR), specific optical rotations, circular dichroism (CD), and gel permeation chromatography (GPC) techniques. The separations of racemates using the CSPs prepared from the polymers were carried out by HPLC.

2.12.1 Monomer ((S)-MLMI)

The monomer was synthesized into 2 step process and the preparation process is described elsewhere [88]. In brief, the reaction mixture of L-leucine, thionyl chloride (SOCl$_2$), methanol (MeOH) at −10 °C was brought to room temperature after stirring for 15 h. After completion of reaction, methanol and excess thionyl chloride were removed by distillation under reduced pressure and the residue was crystallized from diethyl ether. After filtered and dried under vacuum, (S)-leucine methyl ester hydrochloride was obtained as a white powder which was suspended into ethyl acetate (EtOAc), and triethylamine (Et$_3$N) was dropwise added, stirred over 10 min, and collected by suction filtration. Maleic anhydride dissolved into ethyl acetate was added into the filtrate at room temperature, stirred over 24 h. The mixture was washed by H$_2$O, saturated brine, and dried over Na$_2$SO$_4$, and concentrated to obtain (S)-maleamic acid-L-leucine methyl ester ((S)-MLMA) as white powder.

In second step, (S)-MLMA was dissolved in benzene, Celite 545, ZnBr$_2$ were added and heated to 50 °C. After addition of 1,1,1,3,3,3-hexamethyldisilazane (HMDS) solution in benzene the reaction temperature was raised to 80 °C for 12 h with stirring. The solution mixture was filtered and the residue obtained was dissolved into ethyl acetate. It was washed with 0.1 N HCl aq., saturated NaHCO$_3$, and brine and dried over Na$_2$SO$_4$ and concentrated. The crude product was purified by column chromatography and vacuum distillation to obtain (S)-MLMI, yield: 62.9 % [88].

2.12.2 Model Compound (S)-N-Succinoyl-L-Leucine Methyl Ester ((S)-MLSI)

In a Schlenk reaction tube (S)-MLMI was dissolved in ethyl acetate, and 10 % palladium-activated carbon was added to the solution. For evacuation of reaction mixture, the aspirator was used and the same was replaced by hydrogen gas 5 times. After stirred under hydrogen atmosphere for 15 h, the reaction tube was again evacuated by an aspirator and replaced by nitrogen gas. Thereafter the reaction mixture was filtered to remove palladium-activated carbon. Finally, the filtrate was concentrated under pressure to afford (S)-N-succinoyl-L-leucine methyl ester ((S)-MLSI) as a white solid with 99.7 % yield [88].

2.12.3 ((S)-BnLMI) Monomer

Table 1 shows the synthesis scheme of the monomer. The synthesis method is described elsewhere [88]. In brief, L-Leucine, benzyl alcohol and p-toluenesulfonic acid (p-TsOH) in benzene was refluxed for 5 h. The crystallization of the product,

(S)-leucine benzyl ester p-toluenesulfonate was made after concentration of the mixture from diethyl ether with yield: 100 %. Then it was suspended into ethyl acetate and the solution of triethylamine in ethyl acetate was added at room temperature, filtered and then a solution of maleic anhydride in ethyl acetate was added into it with stirring for 24 h. To obtain (S)-N-leucine benzyl ester maleamic acid ((S)-BnLMA) the reaction mixture was washed with water, saturated brine, and dried over Na_2SO_4 and concentrated under reduced pressure. Then (S)-BnLMI was synthesized from (S)-BnLMA in a similar manner as described to (S)-MLMI.

To exclude oxygen and moisture, all experiments for polymerization reactions were carried out under purified nitrogen atmosphere. Anionic polymerizations were carried out in the following procedures [88]. In brief, to dissolve monomer and chiral ligand separately THF or toluene solvent was added by a syringe to each vessel under nitrogen atmosphere. Organometal (n-BuLi, Me_2Zn, or Et_2Zn) in n-hexane solution was introduced into the chiral ligand solution to prepare initiator complex. The complex solution and monomer solution were added under nitrogen atmosphere to initiate polymerization. The polymerization was terminated with a small amount of methanol containing hydrochloric acid. The polymer was precipitated into a large amount of methanol, filtered, washed with methanol, and dried. Re-precipitation of polymer was done in a THF-methanol system, dried under vacuum at room temperature for 2 d. Whereas, radical polymerization was conducted in THF or toluene in a sealed tube at 60 °C for 24 h using AIBN as an initiator. The precipitation and purification of polymer was done in similar manner as followed for anionic polymerization.

Polymerizations: The anionic polymerization route was followed to prepare polymers from (S)-MLMI and (S)-BnLMI monomers respectively [88]. Solubility of obtained polymers was in THF and $CHCl_3$ solution and having the negative specific rotations. The presence of organometals, ligands, and temperature are the factors which affects the number-average molecular weights (M_ns), specific rotations and the yields. In the anionic polymerization of (S)-MLMI with n-BuLi initiator series, the yields of poly((S)-MLMI)s increases in THF and toluene in comparison to n-BuLi, but in M_ns of polymers big changes were not observed. Et_2Zn initiator series having the same tendency. Organometal/ligand complex having higher catalytic activity than the organometals. On decreasing the ratio of ligand to Et_2Zn, at 0 °C the obtained yields of poly((S)-MLMI) with Et_2Zn/ligand (1.0/0.5) both decreased comparing with those obtained with Et_2Zn/ligand (1.0/1.2), while increases the M_ns of polymers. By using the Me_2Zn/ligand shows the same results and below −40 °C this affinity was not occurs in the polymerization (Table 1).

Recently, Mallakpour et al. [95] reported a new class of optically active polymers, poly(amide-thioester-imide)s (PATEI)s, containing flexible and anticorrosion linkages following step growth polymerization of a novel diamine with diacids containing amino acid and trimellitylimide groups using tetrabutylammonium bromide surfactant as molten ionic salt. The obtained polymeric materials were of randomly distributed nanoscale assembled filamentary particles (∼40 nm). The polymers were soluble in common organic solvents and formed low-coloured and flexible thin films. The PATEI films exhibited high-optical transparency and absorption edge wavelength (λ_0). The polymers were characterized by various analysis techniques. The effect of ultrasound on the morphology of polymers claimed that the size of polymer particles decreased during change in its morphology.

3 Characterization Techniques for Analyzing the Optical Activities of Polymers

The characterization of synthesized organic material is very common from analysis and confirmation point of views. There are various characterization techniques FTIR, Raman and NMR which are very common to all kind of materials to determine its composition. Thermal properties such as the glass transition temperature, T_g and melting point can be determined by differential scanning calorimetry (DSC) and dynamic mechanical analysis (DMA). Pyrolysis followed by analysis of the fragments is one more technique for determining the possible structure of the polymer. Thermogravimetry is a useful technique to evaluate the thermal stability of the polymer. Detailed analysis of TG curves also allows knowing a bit of the phase segregation in polymers.

A variety of laboratory techniques are known to determine the properties of polymers. Techniques such as wide angle X-ray scattering, small angle X-ray scattering and small angle neutron scattering are used to determine the crystalline structure of polymers. The study of the crystallization of enantiomers and their packing arrangement in a crystal lattice led to many insights regarding their mutual interactions and elementary properties. Gel permeation chromatography is used to determine the number average molecular weight, weight average molecular weight, and polydispersity index values.

However, the characterization of optically active polymers requires various parameters for the optical activities which need to be specified. This is because a polymer may have a statistical distribution of chains of varying lengths, and monomer residues at the end which affect its properties drastically. Specific optical rotations as one of the chiroptical properties of monomers and polymers are measured on a polarimeter with a 1 dm cell using THF as the solvent. Circular dichroism (CD) spectropolarimeter is used to record spectra using spectrometric grade solvent in 1 and 0.1 cm cells at concentrations of 10^{-4} to 10^{-5} M respectively. The data further needs to be analyzed using the associated software. The textures of monomers and polymers in liquid crystalline form are observed on a hot stage polarizing optical microscope (e.g., Linkam TH-600PM) using solid powder between two glass slides. The digital automatic polarimeter was used for the optical rotation of all the samples at room temperature. The wavelength of the sodium lamp used 589.44 nm. The infrared emissivity values of the samples were investigated.

Birefringence Measurement

Optically active polymer films containing birefringence property may be created by *trans–cis–trans* photoisomerization processes produced with a linear polarized light, with the induced orientation of the chromophores as perpendicular to the polarization direction. The detail of the measurement technique has been described elsewhere by authors' group [96]. The anisotropic molecular orientation results in stable photo induced birefringence or dichroism, are erased by randomization of the alignment with circularly polarized light or heat. For the measurement of the photoinduced birefringence the experimental setup is shown in Fig. 7. For writing 532-nm line from a Nd: YAG laser as a pump beam, and for reading 632.8-nm line from a He–Ne laser as a probe beam were used. A film sample was placed between a pair of crossed polarizers during experiment. The polarization vector of the pump beam used to set at 45 °C to achieve maximum signal with respect to the polarization vector of the probe beam. The quarter-wave plate converted the linear polarized beam to circular polarization, which removed the birefringence that need to be induced.

SHG Measurements

Spin-coating and free standing films are used to measure SHG activity. The circular-difference effects in second-harmonic generation are used to study chiral, anisotropic thin films of a helicene derivative. In general, Q-switched Nd:YAG laser at 1064 nm used as pump beam. The detail of this measurement technique is described elsewhere [97]. In brief, the delay between the flash lamp pump and Q-switch trigger kept at a very large value of 450 ns to decrease the pulse energy (38 mJ) and increase the pulse width (20 ns). A diverging lens of focal length 30 cm generally used to increase the beam diameter to fit the sample diameter. A 90° configuration scheme for SHG measurement is shown in Fig. 8. The scattered second harmonic radiation collected by a lens. The used photo-multiplier tube (PMT) responds the SHG pulse which to be traced through oscilloscope.

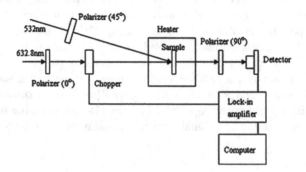

Fig. 7 Experimental setup for birefringence measurement of optically active polymer [96]

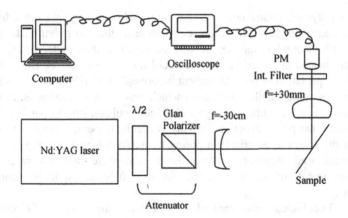

Fig. 8 Experimental set-up for powder SHG of sample [97]

4 Properties

The properties of optically active polymer like its constituent monomers, microstructure and other parameters are very important. The bulk physical properties of the optically active polymers are determined by their basic structures and it also describes behaviours like a continuous macroscopic material, e.g., simultaneous production of L-lactic acid with high optical activity and a soil amendment with food waste that demonstrates plant growth promoting activity [98]. Similarly, the bulk polymer interacts with other chemicals and solvents are described at the macro-scale. Chemical properties, at the nano-scale, describe how the chains interact through various physical forces.

4.1 Nonlinear Optical Properties of Azo-Polymers

The azo polymers show the linear optical properties and nonlinear optical phase conjugation properties. In these polymers the formation of the exciplex by transfer of the electric charges between electron donor and acceptor and shows the large values of molecular hyperpolarizability of the second order. Due to noncentrosymmetric bulk ordering and large second order hyperpolarizability at the molecular level the efficient nonlinear optical properties can only be achieved in materials. NLO properties of azo polymers show due to their good optical characteristics, process property, low cost and wide variety. Consecutively the donor and withdrawing groups of conjugated, synthesized both side chain and main chain of azo polymers and their NLO properties were checked [99–103]. Mostly, amines (–N–) or oxygen (–O–) are used as the electron donor groups and cyano (–CN), nitro (–NO$_2$), or sulfone (–SO$_2$–) are as withdrawing groups.

The azo polymers optical nonlinearities are raised by poling: an electric field was applied at above the glass transition temperature of the azo polymer. In general, between both electrodes, the sample was sandwiched, then under strong electric field, firstly heated at above Tg and then cool down below the Tg (room temperature). The expectation of polar alignment is above Tg and to be frozen at below Tg. In this procedure the stability turned out which was a major drawbacks, in view of the fact that having a rather strong Coulombic repulsion due to parallel dipoles. Consequently, the poled materials stability is relatively poor. Firstly, it was signifying that the increase electric field poling by exploitation of azobenzenes underwent photoinduced alignment. At first, the chromosphere dipoles were created as parallel alignment, in presence of the electrical field and quickly saturate the alignments.

The NLO coefficient was increased ≤ 50 % due to the presence of electric field, after using the polarized light along the same direction of the electric field to favor the alignment of the azobenzene groups. The alignment's relaxation was occurred after turning off the optical pump, but due to the presence of only the electric field, the saturated value was much lower than the final value. The polymer second order NLO properties were obtained when turning off the electric field.

The long time stability of orientation of the azo groups was due to the segmental motion which frozen the chains of the polymer at the lower glass transition temperature. Due to relatively simple and efficient way to improve the electric field poling the alignment was tremendously popular and the method known as "photo assisted" alignment. By Langmuir-Blodgett method, the monomolecular layers at the air-water interface was transferred by mechanically to a solid substrate and obtained in the polar orientation of azo polymers [104]. Without poling process, the alignments of chromophores were permissible in this technique. Most importantly, for the investigation of many nonlinear optical materials and at the fundamental level molecular properties, the LB technique is very powerful. The LB technique having a drawback i.e. a time consuming process, the nonlinear component by the noncontributing aliphatic chains and films tend to be fragile [105–109].

The optical properties (including NLO) of a novel series of pnPEGMAN azopolymers (containing amino-nitro substituted azobenzenes) were studied by absorption spectroscopy and Z-Scan technique. Absorption spectra of the polymer solutions exhibit absorption bands at about λ max = 468–470 nm, and for solid state films λ max = 476–489 nm. The absorption spectrum of p4PEGMAN in solid state film shows the presence of H-aggregates, whereas the rest of the polymers in the solid state only exhibit trace of J-aggregates. The evaluated n_2-values for all the studied materials are in the order of 10^{-4} to 10^{-5} esu, with negative sign. The Z-Scan data inferred that the polymerization of the azo-chromophores and induced *trans-cis* photoisomerization process play major roles in the occurrence of strong cubic nonlinearities. This indicates that the incorporation of PEG-units to the pendant azo-chromophores weakened the NLO-response of polymers as well as decreases the NLO-activity up to one order of magnitude. Hence, the evaluated n_2-values are definitely more related to particular chromophores of polymers for an adequate architecture/functionalization. The addition of PEG-moieties does not play any role (except amphiphilic properties, lowering of Tg-values of pnPEGMAN [109–111].

Azo-based chitosan derivative was prepared by Dutta et al. [97] using different reaction conditions with different characteristic features like crystallinity, good thermal and surface morphological behavior. The optical property measured by UV, AFM and photoluminescence spectra. Second harmonic generation (SHG) of polymers indicated that biopolymers may be considered as potential optical materials in biomedical field of applications.

APUs belongs to the class of film-forming polymeric materials that open wide possibilities for the targeted control of nonlinear-optical characteristics within the same system. This circumstance is primarily related to the possibility to change the chemical structure of the initial hydroxyl containing azo monomers, a process that is primarily directed at improvement of their molecular hyperpolarizability. This problem may be solved through the extension of the conjugated chain length. This can be accomplished for instance via incorporation of additional azobenzene fragments or various heterocyclic systems into chromophores between their electron donor and electron acceptor groups. The same aim may be achieved through variation in the chemical structure of electron-acceptor groups and in their location in azo chromophores, whereas control over the optical properties of the azo-chromophore through a change in the electron-donor part was implemented to a smaller extent. Additional possibilities for optimization of the optical characteristics of the azo-monomer were provided by a change in its molecular architecture. In control of the nonlinear-optical properties of APUs, a special role is played by separation or isolation groups (SIG) that are introduced into the composition of the monomer to stabilize the necessary orientation of azo chromophores in the polymer matrix that is attained during polarization of the material in an external electric field.

The chemical structure and position of the SIG group in the monomer is very important because this group may change the distribution of the electron density between the donor and acceptor parts and, moreover, may possess electron-donor or electron-acceptor properties. In addition, there is a need to optimize the geometry and shape of the SIG group, because it, in many respects, determines the degree of realization of nonlinear-optical characteristics of the initial azo monomer during its incorporation into the polymer structure. The method of incorporating SIG groups into the polymer chain (in the initial azo monomer or via polymer-analogous transformations) makes it possible to additionally broaden the range of variation of the APU's in the nonlinear optical characteristics. The nonlinear-optical characteristics of APUs may be controlled through the use of diisocyanates. However, their commercial assortment is not wide. The incorporation of azo chromophores into side or main chains of a polymer results in a certain difference in their nonlinear-optical characteristics and thus provides another way to attain desired optical properties of the APUs. As a rule, preservation of the necessary temperature and time stability of the nonlinear-optical characteristics of the APUs is related to an increase in their Tg values. This situation may be implemented through incorporation of azo chromophores into the main chain of the polymer or via development of the network structure. Such network APUs show more stable

nonlinear-optical properties; therefore, nonlinear-optical media with a long opera-
tion lifetime may be created on their basis. The APUs are of the interest not only for
the nonlinear-optical applications like second-harmonic generation and
electro-optical modulation considered in this review but also for use as media with
tunable chemical structures and adaptable optical properties. For example, cis–trans
switches observed in many azo benzene fragments at the molecular level may serve
as a basis for the design of "smart surfaces" that can change their properties (e.g.,
the friction coefficient and wettability) or the near-surface orientation of LC
molecules in a desired direction under the action of light of a certain wavelength
and thus can affect the transparency or the index of refraction of thin films [112]. In
addition, chromophore containing polymers have found application as photorefractive nonlinear-optical media, that is, media with tunable refractive indexes and
the capacity for holographic data recording [113]. The cis–trans photoisomerization
of azo chromophores is likewise the key process in surface structuring observed in
some polymers at the nanolevel under the action of light, the so called surface relief
grating effect [114]. Hence, it is safe to say that the APU-based materials belong to
a multifunctional, promising class of polymers. The main problem encountered
during the creation of new nonlinear optical systems based on APUs will be the
synthesis of polymers combining high second order nonlinear optical susceptibility
with good solubility in common organic solvents.

4.2   Thermal Analysis

Polymers used in optoelectronics applications, thermal stability, i.e., decomposition
temperature, $T_d \sim 300$ °C is being considered as one of the most important
properties. The most comprehensive thermal stability of optically active polymers
containing bisazoaromatic chromophores in the side chain are obtained due to the
strong dipolar interactions in the solid state among the side chains.

The thermoplastic poly(imide-amide)s are known for its good heat resistance,
outstanding mechanical properties, and great oxidative stability. These properties
make these polymers most popular in electronic materials, adhesives, composites
materials, fibers, and film materials among others. However, in many cases due to
high rigidity and low solubility in common organic solvents, existence of fully
occupied aromatic rings in the polymer backbone limited its applications. Whereas
the incorporation of chirality factors in the polymers often adds relevant functions
as key basic high-performance materials [115–119]. The synthesis of amino acid
based chiral polymers is an interesting field of research for biomedical applications
due to its harmless, biocompatible, and biodegradable properties.

The photochromic properties of azobenzene derivatives are more pronounced in
chromophoric systems. Azobenzene derivatives when irradiated thermally or by
appropriate light, the more stable azo-trans form isomerizes reversibly to the azo-cis
form and thus possessing higher dipole moment and free volume requirement with

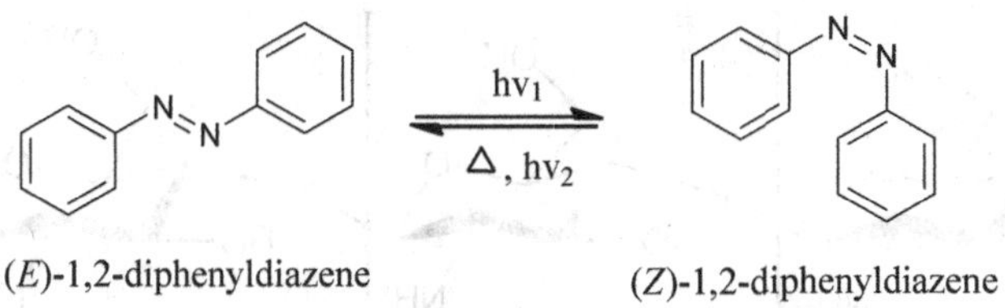

(E)-1,2-diphenyldiazene   (Z)-1,2-diphenyldiazene

Fig. 9 Photoisomerization of azobenzene

several relevant consequences (Fig. 9). The moiety when incorporated into polymers or other materials show fantastic photochromic properties.

Consequent upon these systems into materials may show effects on various chemical, mechanical, electronic and optical properties of the materials and can be used as photoswitches. An exhaustive research work has been carried out by various group of researchers onto the photoinduced motions in azo-containing polymers and liquid crystals as light-responsive materials [120–124].

4.3 Chiroptical Properties

The chiral groups substituted p-conjugated polyacetylenes form helical structure with a predominantly one-handed screw sense exhibits simultaneous changes in both optical and chiroptical properties [125, 126]. For example, the pendant chiral oxazoline moiety in poly(phenylacetylene) [127] induces an excess one handedness helical conformation in the main chain of the polymer.

Optical properties of chitosan/myristic acid and chitosan/nicotinic acid derivatives show red shift. Chitosan/methoxycinnamaldehyde, the N-substituted chitosan derivative showed fairly good photoluminescence (PL) properties, and introduces polymer conformations in organic and inorganic solvents (Fig. 10). The chiro-optical properties of chitosan-derivatives may draw the attention to biomedical applications [128, 129].

4.4 Vapochromic Behaviour

Optically active materials also show the vapochromic properties and developed the volatile-organic-compound (VOC) fiber-optic sensor. The best examples of such properties are Krogmann's salt $K_2[Pt(CN)_4X_{0.3}]$ (platinum metal complex) [130–133], 2,2'-bipyridine and $\{Au_2Ag_2(C_6F_5)_4[(C_5H_4N)]–(C_5H_4N)]_2\}_n$ (gold–silver based complex) [134]. This sensor can easily detect the presence of the vapors of VOCs like ethanol, methanol, acetic acid, ethyl acetate, glycol ethers, acetone etc. Due to its easily accomplishment, low cost and uses in sensor network/telecommunication

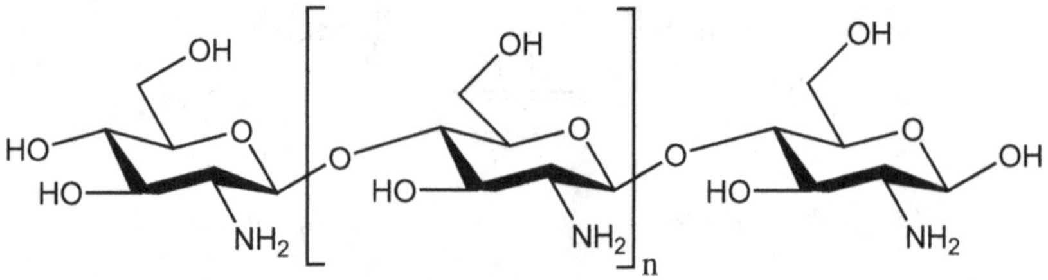

Chitosan

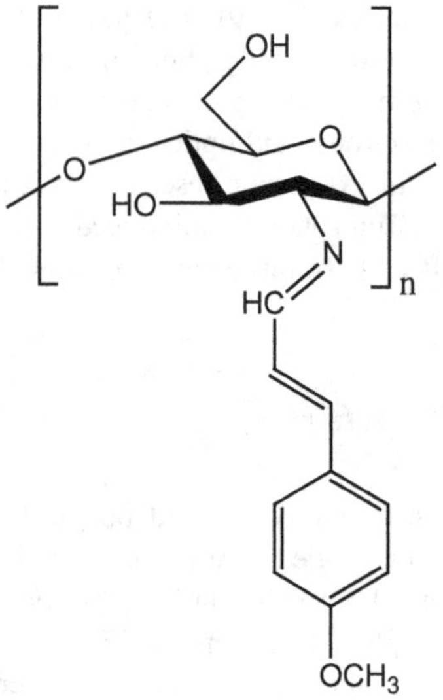

Fig. 10 Structure of chitosan and its derivative

this sensor suitable for use in environmental applications, chemical industries and optoelectronic noses.

4.5 Absorption and Emission Properties

Optically active polymers possess both the absorption and emission properties. It was easily shown in the β-cyclodextrin (β-CD) fluoranthene, benzo[k]fluoranthene and their derivatives, poly(BnEPhOx)s along with BnEPhOx etc. Generally such type of absorption for all polymers occurred at 270–520 nm. In the BnEPhOx, the absorption was shown due to the presence of oxazoline group at 270–330 nm. The spectral data was compared by its corresponding monomer. Such type of polymers show the longer wavelength absorption but not show in its monomer and this occurs due to the conjugated polyacetylene p–p^* interband transition [135, 136].

4.6 Thermosensitivity

Thermosensitive, and amphiphilic polymer brushes optically active polymers consist of helical poly(N-propargylamide) main chains and thermosensitive poly(N-isopropylacrylamide) (PNIPAm) side chains, were prepared via a novel methodology combining catalytic polymerization, atom transfer radical polymerization (ATRP), and click chemistry. The characterization of GPC, FT-IR, and ^{1}H-NMR measurements indicated the successful synthesis of the novel amphiphilic polymer brushes. For the confirmation of helical structure of the polymers backbones and the optical activity of the final brushes used UV-Vis and CD spectra. The polymer with optically active cores (helical polyacetylenes) and thermosensitive shells (PNIPAm) brushes self-assembled in aqueous solution forming core/shell structured nanoparticles [137].

Optically active polymers show another properties namely thermosensitivity, e.g., main chains helical poly(N-isopropylacrylamide) and thermosensitive part as side chain of poly(N-isopropylacrylamide) (PNIPAm). Such type of polymers synthetic method described elsewhere [137]. The polymer with optically active cores (helical polyacetylenes) and thermosensitive shells (PNIPAm) brushes self-assembled core/shell structured nanoparticles in aqueous solution. Another example of optically active polymer is poly[N-(L)-(1-hydroxymethyl)-propylmethacrylamide] (P(L-HMPMA)) of lower critical solution temperature and thermosensitivity. Circular dichroism and microcalorimetric measurements of the polymer showed the polymer chains in a state of relatively low hydration compared to that of by racemate synthesized monomers by free-radical reaction formed P(D,L-HMPMA). Thermosensitivity and structural effects were obtained by microscopic observation of aqueous solution of polymers and its hydrogels [138].

4.7 Chiral Separation

Chiral separation is an analytical technique for evaluation of enantiomeric purity of chiral compounds. Such property also found in the optically active polymers. Polyelectrolyte "multilayer" (PEMU) is an example for such property and made up by polypeptides, such as L-and D-poly(lysine), poly(glutamic acid), poly(N-(S)-2-methylbutyl-4-vinylpyridininum iodide), poly(styrene sulfonate) etc. PEMUs allow at very high enantiomer permeation rates for chiral membrane separations [139–141].

4.8 Fabrication

The fabrication of three-dimensional microstructures are made by the polymerization of two-photons, and the process described elsewhere [142–144]. In brief, 2-methoxy-5-2-ethylhexyloxy-1,4-phenylenevinylene (MEH-PPV) used for two-photon polymerization of microstructures. MEH-PPV polymer is guest material having conductivity, electroluminescence, and nonlinear optical properties. An acrylate-based resin acts a matrix for fabrication of microstructures with MEH-PPV by two-photon absorption polymerization. The resin contains two triacrylate monomers namely tris 2-hydroxyethyl isocyanurate triacrylate and ethoxylated trimethylolpropane triacrylate. One increases the hardness and another reduces shrinkage upon polymerization. The microstructures fabricated, with an average power of 40 mW and laser beam focused into sample, laser fluence yielding of nearly 30 mJ/cm^2 [145].

4.9　Photochromic Property

Optically active compounds show photochromic property when irradiated with suitable frequency and intensity of light, for example, waveguides, optical switches, photochromic, optical modulators, surface relief gratings etc. This is due to the electronic transitions of the azo group, the photochromic transazoaromatic groups undergo a reversible and repeated trans-cis-trans isomerization [146]. A few compounds are reported dealing with optically active synthetic polymers with hydrocarbon main chains and side-chain azobenzene moieties. Poly (arylene-ethynylene)-alt-poly(arylene-vinylene) (PAE-PAV) is an example for such property. It contains an anthracene and statutory unit (–Ph–C≡C–Anthr–C≡C–Ph–CH=CH–Anthr–CH=CH)$_n$ containing two 2-ethylhexyloxy side chains on each phenylene (Ph) unit [147–151]. After synthesis and characterization studied the properties show its photochromic property. Polymeric derivatives of poly (phenylene vinylene)s, poly(phenylene ethynylene)s and polythiophenes are used as semiconducting materials in optoelectronics [152–154].

5　Applications

Optically active polymers (OAPs) have made many significant contributions to optical fibers for cable TV, optical scanners, optical fiber telephone cables, optical data storage, optical monitors for antilock brakes, optical fiber dashboard displays, computer-generated optical elements, optical inspection of labeling and packaging, optical stereo-lithography and many more.

Actually, the organic electro-optic materials have potential into commercial devices on bandwidth and information density fields such as information and communication technology, biomedicine and computing which can satisfy different sectors like biomedical imaging, high speed communications, electronics, etc. Further, OAPs possess photonics, which includes ultrafast all-optical modulation in polymer photonic crystals, silicon/organic hybrid systems, gain switching in polymer amplifiers and lasers, and new devices such as hybrid organic/inorganic electrically pumped lasers have the potential to meet the ever-increasing demands of our modern society.

OAPs definitely play important role as catalysts, particularly, in asymmetric synthesis and separation of recemic mixtures. These are of particular interest due to their specific properties and successful uses as stationary phase in chromatographic separation of enantiomers, chiral media for asymmetric synthesis, chiral liquid crystals in ferroelectric and nonlinear optical devices/storage, optical switches, biodegradable and biomedical devices, drug delivery agents, to photorefractive and photoconductive applications etc. The condensed packing problem existing in the dendrimer-based chiral catalysts without sacrificing the advantage of easy recovery by either filtration or precipitation could be overcome by using dendronized optically active polymers (lower generation dendritic wedges) with monomeric chiral units in their peripheries.

As a conducting polymer, it has the potential of combining the high conductivity normally associated with the metals and the processibility, corrosion resistance and low density of organic polymers and also used as battery materials, electro chromic displays and sensors. The fruition of OAPs in terms of its higher ordered structures of macromolecules would be essential in their function, including molecular recognition, catalytic activity and substrate specificity with an exciting present and promising future.

6   Conclusions

From the above discussion it is clear that the optically active multifunctional polymeric derivatives offer a convenient route to synthesize various kinds of polymers/copolymers with optical activity, intensity of dichroic bands, amplification of chirality, achieving of high Tg values for efficient applications in optoelectronics, storage capacity for optical information, waveguides, chiroptical switches, chemical photoreceptors, NLO, surface relief gratings (SRG), photoconductive properties and many more. So the need for a systematic study on syntheses, characterization and properties of optically active polymers has been understood. In this monograph various aspects including the synthesis, reaction conditions and properties have been critically emphasized and such studies would appear to be interesting and useful for various applications.

References

1. T. Nakano, Y. Okamoto, K. Hatada, Asymmetric polymerization of triphenylmethyl methacrylate leading to a one-handed helical polymer: mechanism of polymerization. J. Am. Chem. Soc. **114**, 1318–1329 (1992)
2. J.S. Moore, S.I. Stupp, Materials chemistry of chiral macromolecules. 1. Synthesis and phase transitions. J. Am. Chem. Soc. **114**, 3429–3441 (1992)

4. K.L. Singfield, G.R. Brown, Optically active polyethers. 1. Studies of the crystallization in blends of the enantiomers and the stereoblock form of Poly(epichlorohydrin). Macromolecules **28**, 1290–1297 (1995)

5. H. Schlaad, H. Kukula, B. Smarsly, M. Antonietti, T. Pakula, Solid-state morphologies of linear and bottlebrush-shaped polystyrene—poly(Z-L-lysine) block copolymers. Polymer **43**, 5321–5328 (2002)

6. J. Pecher, S. Mecking, Nanoparticles from step-growth coordination Polymerization. Macromolecules **40**, 7733–7735 (2007)

7. B. Zhao, J. Deng, J. Deng, Emulsification-Induced homohelicity in racemic helical polymer for preparing optically active helical polymer nanoparticles. Macromol. Rapid Commun. **37**, 568–574 (2016)

8. R. Wang, Y. Zheng, X. Li, J. Chen, J. Cui, J. Zhang, X. Wan, Optically active helical vinylbiphenyl polymers with reversible thermally induced stereomutation. Polym. Chem. **7**, 3134–3144 (2016)

9. Y. Miyagi, T. Hirao, T. Haino, F. Sanda, Synthesis of optically active conjugated polymers containing platinum in the main chain: control of the higher-order structures by substituents and solvents. J. Poly. Sci. Part A Poly. Chem. **53**, 2452–2461 (2015)

10. H. Huang, C. Chen, D. Zhang, J. Deng, Y. Wu, Helical substituted polyacetylene derived fluorescent microparticles prepared by precipitation polymerization. Macromol. Rapid Commun. **35**, 908–915 (2014)

11. C. Song, X. Liu, D. Liu, C. Ren, W. Yang, J. Deng, Optically active particles of chiral polymers **34,** 1426–1445 (2013)

12. J. Liu, J.W.Y. Lam, B.Z. Tang, Acetylenic polymers: syntheses, structures, and functions. Chem. Rev. **109**, 5799–5867 (2009)

13. C.S. Daeffler, G.M. Miyake, J. Li, R.H. Grubbs, Partial kinetic resolution of oxanorbornenes by ring opening metathesis polymerization with a chiral ruthenium initiator. ACS Macro Lett. **3**, 102–104 (2014)

14. E. Yashima, K. Maeda, H. Iida, Y. Furusho, K. Nagai, Helical polymers: Synthesis, structures and functions. Chem. Rev. **109**, 6102–6211 (2009)

15. Z.B. Zhang, M. Motonaga, M. Fujiki, C.E. McKenna, The first optically active polycarbazoles. Macromolecules **36**, 6956–6958 (2003)

16. S. Jin, T.H. Tiefel, R. Wolfe, R.C. Sherwood, J.J. Mottine Jr., Optically transparent, electrically conductive composite medium. Science **255**, 446–448 (1992)

17. H.S. Kim, J.E. Park, M.K. Patel, H. Kim, D.S. Kim, S.K. Byeon, D. Lim, J. Kim, Optically transparent and electrically conductive NiO window layer for Si solar cells. Mat. Lett. **174**, 10–13 (2016)

18. L. Chen, Y. Chen, K. Yao, W. Zhou, F. Li, L. Chen, R. Hu, B.Z. Tang, A novel type of optically active helical liquid crystalline polymers: synthesis and characterization of poly (p-phenylene)s containing terphenyl mesogen with different terminal groups. J. Poly. Sci. Part A Poly. Chem. **47**, 4723–4735 (2009)

19. Y. Suzuki, M. Shiotsuki, F. Sanda, T. Masuda, Chiral 1-methylpropargyl alcohol: a simple and powerful helical source for substituted polyacetylenes. Macromolecules **40**, 1864–1867 (2007)

20. T. Aoki, T. Kaneko, N. Maruyama, A. Sumi, M. Takahashi, T. Sato, M. Teraguchi, Helix-sense-selective polymerization of phenylacetylene having two hydroxy groups using a chiral catalytic system. J. Am. Chem. Soc. **125**, 6346–6347 (2003)

21. B.S. Li, K.K.L. Cheuk, L. Ling, J. Chen, X. Xiao, C. Bai, B.Z. Tang, Synthesis and hierarchical structures of amphiphilic polyphenylacetylenes carrying L-valine pendants. Macromolecules **36**, 77–85 (2003)

22. J.W.Y. Lam, B.Z. Tang, Functional polyacetylenes. Accounts Chem. Res. **38**, 745–754 (2005)

23. K. Okoshi, S.I. Sakurai, S. Ohsawa, J. Kumaki, E. Yashima, Control of main-chain stiffness of a helical poly(phenylacetylene) by switching on and off the intramolecular hydrogen bonding through macromolecular helicity inversion. Angew. Chem. Int. Ed. **48**, 8173–8176 (2006)

24. S. Wu, N. Yang, L. Yang, J. Cao, J. Liu, A novel type of optically active helical polymers: synthesis and characterization of poly(-unsaturated ketone). J. Poly. Sci. Part A Poly. Chem. **48**, 1441–1448 (2010)

25. S. Zahmatkesh, M.R. Vakili, Synthesis and characterization of new optically active poly (ethyl L-lysinamide)s and poly (ethyl L-lysinimide)s. J. Amino Acids **2010**, 1–6 (2010)

26. S.E. Mallakpour, A.R. Hajipour, S. Khoee, optically active poly(amide–imide)s by direct polycondensation of aromatic dicarboxylic acid with aromatic diamines. Europ. Poly. J. **38**, 2011–2016 (2002)

27. S.E. Mallakpour, A.R. Hajipour, K. Faghihi, Microwave-assisted synthesis of optically active poly(amide-imide)s with benzophenone and L-alanine linkages. Europ. Poly. J. **37**, 119–124 (2001)

28. K. Faghihi, K. Zamani, A. Mirsamie, M.R. Sangi, Microwave-assisted rapid synthesis of novel optically active poly(amide-imide)s containing hydantoins and thiohydantoins in main chain. Europ. Poly. J. **39**, 247–254 (2003)

29. L.I. Subbotina, A.A. Bakanova, E.R. Kofanov, E.N. Popova, E.N. Vlasova, V.M. Svetlichnyi, optically active polyamidoimides based on amino acids containing cyclohexane fragment. Rus. J. Appl. Chem. **88**, 1661–1666 (2015)

30. P. Rattanatraicharoen, K. Shintaku, K. Yamabuki, T. Oishi, K. Onimura, Synthesis and chiroptical properties of helical poly(phenylacetylene) bearing optically active chiral oxazoline Pendants. Polymer **53**, 2567–2573 (2012)

31. Q.S. Hu, C. Sun, C.E. Monaghan, Optically active dendronized polymers as a new type of macromolecular chiral catalysts for asymmetric catalysis. Tetrahed. Lett. **43**, 927–930 (2002)

32. J. Luo, M. Haller, H. Li, H.-Z. Tang, A.K.-Y. Jen, K. Jakka, C.-H. Chou, C.-F. Shu, A side-chain dendronized nonlinear optical polyimide with large and thermally stable electrooptic activity. Macromolecules **37**, 248–250 (2004)

33. S. Mallakpour, S. Habibi, Microwave-promoted synthesis of new optically active poly (ester-imide)s derived from N, N'-(pyromellitoyl)-bis-L-leucine diacid chloride and aromatic diols. Europ. Poly. J. **39**, 1823–1829 (2003)

34. F. Andreani, L. Angiolini, V. Grenci, E. Salatelli, Optically active polyalkylthiophenes: synthesis and polymerization of chiral, symmetrically substituted, quinquethiophene monomer. Synth. Metals **145**, 221–227 (2004)

35. R.H. Mitchell, Y.H. Lai, R.V. Williams, N-Bromosuccinimide-dimethylformamide: a mild, selective nuclear monobromination reagent for reactive aromatic compounds. J. Org. Chem. **44**, 4733–4737 (1979)

36. K. Tamao, S. Kodama, I. Nakajima, M. Kumada, A. Minato, K. Suzuki, Nickel-phosphine complex-catalyzed Grignard coupling—II: grignard coupling of heterocyclic compounds. Tetrahedron **38**, 3347–3354 (1982)

37. F. Andreani, L. Angiolini, D. Caretti, E. Salatelli, Synthesis and polymerization of 3,3″-di [(S)-(+)-2-methylbutyl]-2,2′:5′,2″-terthiophene: a new monomer precursor to chiral regioregular poly(thiophene). J. Mater. Chem. **8**, 1109–1111 (1998)

38. A. Natansohn, in *Symposium on Azobenzene-Containing Materials (1998:) Proceedings of the symposium on azobenzene-containing materials*. Macromolecular Symposia, vol. 137, 1–165 (Boston, MA, 1998)

39. A. Natansohn, P. Rochon, M.S. Ho, C. Barrett, Azo polymers for reversible optical storage. 6. Poly[4-[2-(methacryloyloxy)ethyl]azobenzene]. Macromolecules **28**, 4179–4183 (1995)

40. Y. Wu, Q. Zhang, A. Kanazawa, T. Shiono, T. Ikeda, Y. Nagase, Photo induced alignment of polymer liquid crystals containing azobenzene moieties in the side chain. 5. Effect of the azo contents on alignment behavior and enhanced response. Macromolecules **32**, 3951–3956 (1999)

41. C.R. Mendonca, A. Dhanabalan, D.T. Balogh, L. Misoguti, D.S.D. Santos Jr., M.A. Pereira-da-Silva, J.A. Giacometti, S.C. Zilio, O.N. Oliveira Jr., Optically induced birefringence and surface relief gratings in composite langmuir—blodgett (LB) films of poly[4-[[2-(methacryloyloxy)ethyl]ethylamino]-2-chloro-4-nitroazobenzene](HPDR13) and cadmium stearate. Macromolecules **32**, 1493–1499 (1999)
42. K. Ichimura, Photoalignment of liquid-crystal systems. Chem. Rev. **100**, 1847–1874 (2000)
43. C. Zhao, K. Ouyang, N. Yang, J. Zhang, Z. Yang, Synthesis and properties of optically active helical polyethers bearing indole or carbazole derivatives. Macromol. Res. **24**, 393–399 (2016)
44. L. Angiolini, L. Giorgini, H. Li, A. Golemme, F. Mauriello, R. Termine, Synthesis, characterization and photoconductive properties of optically active methacrylic polymers bearing side-chain 9-phenylcarbazole moieties. Polymer **51**, 368–377 (2010)
45. S. Mallakpour, A. Zadehnazari, Advances in synthetic optically active condensation polymers—A review, eXPRESS. Poly. Lett. **5**, 142–181 (2011)
46. A. Natansohn, P. Rochon, Photoinduced motions in azo-containing polymers. Chem. Rev. **102**, 4139–4175 (2002)
47. L. Li, H.C. Dong, Y. Zhang, Z.D. Xu, X.H. Fan, X.F. Chen, Q.F. Zhou, Photoinduced holographic phase grating buried in azobenzene side-chain polymer films with a chiral group. Chinese J. Polym. Sci. **21**, 93–98 (2003)
48. T. Verbiest, M. Kauranen, A. Persoon, Second-order nonlinear optical properties of chiral thin films. J. Mat. Chem. **9**, 2005–2012 (1999)
49. S.A. Kandjani, R. Barille, J.M. Nunzi, R. Kheradmand, H. Tajalli, Light induced 2D chiral structure on the surface of azo-polymer films. Phys. Stat. Solid **8**, 2773–2776 (2011)
50. A. Apostoluk, J.M. Nunzi, C. Fiorini-Debuisschert, Photo-induction of surface relief gratings during all optical poling of polymer films. Opt. Lett. **29**, 98–100 (2004)
51. Y. Wu, A. Natansohn, P. Rochon, Photoinduced birefringence and surface relief gratings in polyurethane elastomers with azobenzene chromophore in the hard segment. Macromolecules **37**, 6090–6095 (2004)
52. L. Angiolini, R. Bozio, A. Dauru, L. Giorgini, D. Pedron, G. Turco, Photomodulation of the chiroptical properties of new chiral methacrylic polymers with side chain azobenzene moieties. Chem. Eur. J. **8**, 4241–4247 (2002)
53. K.A. Gunay, N. Schuwer, H.A. Klok, Synthesis and post-polymerization modification of poly (pentafluorophenyl methacrylate) brushes. Polym. Chem. **3**, 2186–2192 (2012)
54. A.J. Wilson, M. Masuda, R.P. Sijbesma, E.W. Meijer, Chiral amplification in the transcription of supramolecular helicity into a polymer backbone. Chem. Int. Ed. **44**, 2275–2279 (2005)
55. L. Angiolini, R. Bozio, T. Dainese, L. Giorgini, A. Golemme, F. Mauriello, D. Pedron, R. Termine, Photoresponsive polymers containing side-chain chiral azocarbazole chromophores as multifunctional materials. Proc. SPIE **6653**, 665305 (2007)
56. L. Angiolini, T. Benelli, L. Giorgini, F. Mauriello, E. Salatelli, Chiroptical and thermoplastic acid sensors based on chiral methacrylic polymers containing azoaromatic moieties. Sens. Actuators B **126**, 56–61 (2007)
57. E. Sperotto, G.P.M.V. Klink, G.V. Koten, J.G. de Vries, The mechanism of the modified Ullmann reaction. Dalton Trans. **39**, 10338–10351 (2010)
58. D.W. Kim, H. Moon, S.Y. Park, S.I. Hong, Synthesis of photoconducting nonlinear optical side-chain polymers containing carbazole derivatives. React. Funct. Poly. **42**, 73–86 (1999)
59. L. Angiolini, T. Benelli. R. Bozio, A. Dauru, L. Giorgini, D. Pedron, E. Salatelli, Improvement of photoinduced birefringence properties of optically active methacrylic polymers through copolymerization of monomers bearing azoaromatic moieties. Macromolecules **39**, 489–497 (2006)
60. C Carlini, L. Angiolini, D. Caretti, *Photochromic Optically Active Polymers in Polymeric Materials Encyclopaedia*. editor-in-chief J.C. Salomone (CRC Press, Boca Raton, 1996), vol. **7**, pp 5116–5123

61. I.V. Taydakov, S.A. Ambrozevich, E.A. Varaksina, A.G. Vitukhnovsky, A.A. Tyutyunov, O.A. Melnik, Luminescent properties of a composite of acrylic polymers doped with Eu(III) complex for ink-jet printing applications. J. Russ. Laser Res. **37**, 192–196 (2016)

62. L. Angiolini, T. Benelli, L. Giorgini, Synthesis and chiroptical properties of chiral azoaromatic dendrimers with a C₃-symmetrical core. Chirality **22**, 99–109 (2010)

63. L. Angiolini, D. Caretti, L. Giorgini, E. Salatelli, Synthesis and chiroptical properties of optically active methacrylic polymers bearing the (*S*)- and/or (*R*)-2-hydroxysuccinimide moiety linked to the *trans*-azobenzene group in the side chain. Macromol. Chem. Phys. **201**, 533–542 (2000)

64. L. Angiolini, T. Benelli, L. Giorgini, A. Painelli, F. Terenziani, Chiral interactions in azobenzene dimers: a combined experimental and theoretical study. Chem. Eur. J. **11**, 6053–6353 (2005)

65. L. Angiolini, T. Benelli, L. Giorgini, E. Salatelli, R. Bozio, A. Dauru, D. Pedron, Synthesis, chiroptical properties and photoinduced linear birefringence of the homopolymer of (R)-3-methacryloyloxy-1-(4'-cyano-4-azobenzene) pyrrolidine and of the copolymers with the enantiomeric monomer. Europ. Poly. J. **41**, 2045–2054 (2005)

66. M. Ho, C. Barrett, J. Paterson, M. Esteghamatian, A. Natansohn, P. Rochon, Synthesis and optical properties of poly{(4-nitrophenyl)-[3-[*N*-[2-(methacryloyloxy) ethyl]- carbazolyl]]-diazene}. Macromolecules **29**, 4613–4618 (1996)

67. V.A. Miller, R.R. Brown, E.B. Gienger Jr., US Patent 3 067 180 (1962)

68. J.S. Moore, S.I. Stupp, Room temperature polyesterification. Macromolecules **23**, 65–70 (1990)

69. L. Angiolini, T. Benelli, L. Giorgini, A. Golemme, F. Mauriello, E. Salatelli, R. Termine, Methacrylic polymers containing optically active side-chain carbazole: synthesis, characterization and photoconductive properties. Macromol. Chem. Phys. **209**, 944–956 (2008)

70. H. Katsumi, M. Nishikawa, F. Yamashita, M. Hashida, Development of polyethylene glycol-conjugated Poly-S-Nitrosated serum albumin, a novel S-Nitrosothiol for prolonged delivery of nitric oxide in the blood circulation *In Vivo*. J. Pharmacol. Exp. Therap. **314**, 1117–1124 (2005)

71. S. Wen, X. Bao, W. Shen, C. Gu, Z. Du, L. Han, D. Zhu, R. Yang, Synthesis of benzodithiophene based poly(aryleneethynylene)s: synthesis, optical properties and applications in organic solar cell. J. Poly. Sci. Part A: Poly. Chem. **52**, 208–215 (2014)

72. S.P. Webster, J.R. Seckl, B. Walker, P.R. Ward, T.D. Pallin, H.J. Dyke, T.R. Perrior, PCT Intl WO/2009/074789 (2009)

73. Q. Shi, H. Fan, Y. Liu, W. Hu, Y. Li, X. Zhan, A copolymer of benzodithiophene with TIPS side chains for enhanced photovoltaic performance. Macromolecules **44**, 9173–9179 (2011)

74. H. Ogoshi, T. Mizutani, Multifunctional and chiral porphyrins: model receptors for chiral recognition Acc. Chem. Res. **31**, 81–89 (1998)

75. X. Huang, K. Nakanishi, N. Berova, Porphyrins and metalloporphyrins: versatile circular dichroic reporter groups for structural studies. Chirality **12**, 237–255 (2000)

76. X. Huang, B.H. Richman, B. Borhan, N. Berova, K. Nakanishi, Zinc porphyrin tweezer in hosteguest complexation: determination of Absolute configurations of diamines, amino acids and amino alcohols by circular dichroism. J. Am. Chem. Soc. **120**, 6185–6196 (1998)

77. G. Proni, G. Pescitelli, X. Huang, K. Nakanishi, N. Berova, Magnesium tetraaryl porphyrin tweezer: a CD-sensitive host for absolute configurational assignments of achiral carboxylic acids. J. Am. Chem. Soc. **125**, 12914–12927 (2003)

78. T. Kurtan, N. Nesnas, Y. Li, X. Huang, K. Nakanishi, N. Berova, Chiral recognition by CD-sensitive dimeric zinc porphyrin host. 1. Chiroptical protocol for absolute configurational assignments of monoalcohols and primary monoamines. J. Am. Chem. Soc. **123**, 5962–5973 (2001)

79. J. Lu, L. Wu, L. Jing, X. Xu, X. Zhang, Synthesis, circular dichroism, and third-order nonlinear optical properties of optically active porphyrin derivatives bearing four chiral citronellal moieties. Dyes Pigm. **94**, 169–174 (2012)

80. V.V. Borovkov, J.M. Lintuluoto, M. Sugiura, Y. Inoue, R. Kuroda, Remarkable stability and enhanced optical activity of a chiral supramolecular bisporphyrin tweezer in both solution and solid state. J. Am. Chem. Soc. **124**, 11282–11283 (2002)

81. S. Tamaru, M. Takeuchi, M. Sano, S. Shinkai, Sol-gel transcription of sugar appended porphyrin assemblies into fibrous silica: unimolecular stacks versus helical bundles as templates. Angew. Chem. Int. Ed. **41**, 853–856 (2002)

82. S.I. Tamaru, M. Nakamura, M. Takeuchi, S. Shinkai, Rational design of a sugar appended porphyrin gelator that is forced to assemble into a one dimensional aggregate. Org. Lett. **3**, 3631–3634 (2001)

83. X.M. Guo, C. Jiang, T.S. Shi, Prepared chiral nanorods of a cobalt (II) porphyrin dimer and studied changes of UV-Vis and CD spectra with aggregate morphologies under different temperatures. Inorg. Chem. **46**, 4766–4768 (2007)

84. I.T. Ishi, J.H. Jung, S. Shinkai, Intermolecular porphyrinefullerene interactioncan reinforce the organogel structure of a porphyrin-appended cholesterol derivative. J. Mater. Chem. **10**, 2238–2240 (2000)

85. S. Tamaru, S. Uchino, M. Takeuchi, M. Ikeda, T. Hatano, S. Shinkai, A porphyrin based gelator assembly which is reinforced by peripheral urea groups and chirally twisted by chiral urea additives. Tetrahedron Lett. **43**, 3751–3755 (2002)

86. J. Lu, L. Wu, J. Jiang, X. Zhang, Helical nanostructures of an optically active metal free porphyrin with four optically active binaphthyl moieties: effect of metaleligand coordination on the morphology. Eur. J. Inorg. Chem. **25**, 4000–4008 (2010)

87. J. Lu, L. Wu, L. Jing, X. Xu, X. Zhang, Synthesis, circular dichroism and third-order nonlinear optical properties of optically active porphyrin derivatives bearing four chiral citronellal moieties. Dyes Pigm. **94**, 169–174 (2012)

88. T. Oishi, H. Gao, T. Nakamura, Y. Isobe K. Onimura, Asymmetric polymerizations of N-substituted maleimides bearing L-leucine ester derivatives and chiral recognition abilities of their polymers. Poly. J. **39**, 1047–1059 (2007)

89. T. Oishi, K. Onimura, W. Sumida, T. Koyanagi, and H. Tsutsumi Asymmetric anionic polymerization of n-diphenyl-methylitaconimide with chiral ligand-organometal complex. Poly. Bull. **48**, 317–325 (2002)

90. Y. Isobe, H. Tsutsumi, T. Oishi, Asymmetric polymerization of N-1-Naphthylmaleimide with chiral anionic initiator: preparation of highly optically active poly (N-1-naphthylmaleimide). Macromolecules **34**, 7617–7623 (2001)

91. Y. Isobe, K. Onimura, H. Tsutsumi, T. Oishi, Asymmetric anionic polymerization of N-1-naphthylmaleimide with chiral ligand-organometal complexes in toluene. J. Poly. Sci. Part A Poly. Chem. **39**, 3556–3565 (2001)

92. K. Onimura, Y. Zhang, M. Yagyu, T. Oishi, Asymmetric anionic polymerization of optically active N-1-cyclohexylethylmaleimide. J. Poly. Sci. Part A Poly. Chem. **42**, 4682–4692 (2004)

93. H. Gao, Y. Isobe, K. Onimura, T. Oishi, Synthesis and polymerization of novel (S)-N-maleoyl-L-leucine propargyl ester. Poly. J. **38**, 1288–1291 (2006)

94. H. Gao, Y. Isobe, K. Onimura, T. Oishi, Synthesis and asymmetric polymerization of (S)-N-maleoyl-L-leucine propargyl ester. J. Poly. Sci. Part A Poly. Chem. **45**, 3722–3738 (2007)

95. S. Mallakpour, A. Zadehnazari, Novel optically active poly(amide-thioester-imide)s containing L-α-amino acids and thiadiazol anticorrosion group: production and characterization. High Perform. Poly. **25**, 377–386 (2012)

96. N. Nigam, S. Kumar, P.K. Dutta, S. Pei, T. Ghosh, Chitosan containing azo-based Schiff bases: thermal, antibacterial and birefringence properties for bio-optical devices. RSC Adv. **6**, 5575–5581 (2016)

97. P.K. Dutta, S. Kumar, N. Nigam, T. Ghosh, S.P. Singh, P.K. Datta, L. An, T.F. Shi, Preparation, characterization and optical properties of a novel azo-based chitosan biopolymer. Mater. Chem. Phys. **120**, 361–370 (2010)

98. V. Kitpreechavanich, A. Hayami, A. Talek, C.F.S. Chin, Y. Tashiro, K. Sakai, Simultaneous production of:L-lactic acid with high optical activity and a soil amendment with food waste that demonstrates plant growth promoting activity. J. Biosci. Bioeng. **122**, 105–110 (2016)

99. S.A. Baeurle, Multiscale modeling of polymer materials using field-theoretic methodologies: a survey about recent developments. J. Math. Chem. **46**, 363–426 (2009)

100. V.V. Shevchenko, A.V. Sidorenko, V.N. Bliznyuk, I.M. Tkachenko, O.V. Shekera, Azo-containing polyurethanes with nonlinear-optical properties. Poly. Sci. Ser. A **55**, 1–31 (2013)

101. A. Altomare, F. Ciardelli, L. Mellini, R. Solaro, Photoactive azobenzene polymers containing carbazole chromophores. Macromol. Chem. Phys. **205**, 1611–1619 (2004)

102. C. Engels, D.V. Steenwinckel, E. Hendrickx, M. Schaerlaekens, A. Persoons, C. Samyn, Efficient fully functionalized photorefractive polymethacrylates with infrared sensitivity and different spacer lengths. J. Mater. Chem. **12**, 951–957 (2002)

103. J. Hwang, J. Sohn, S.Y. Park, Synthesis and structural effect of multifunctional photorefractive polymers containing monolithic chromophores. Macromolecules **36**, 7970–7976 (2003)

104. J. Shi, Z. Jiang, S. Cao, Synthesis of carbazole-based photorefractive polymers via post-azo-coupling reaction. React. Funct. Polym. **59**, 87–91 (2004)

105. T. Kajitani, K. Okoshi, S. Sakurai, J. Kumaki, E. Yashima, Helix-sense controlled polymerization of a single phenyl isocyanide enantiomer leading to diastereomeric helical polyisocyanides with opposite helix-sense and cholesteric liquid crystals with opposite twist-sense. J. Am. Chem. Soc. **128**, 708–709 (2006)

106. E. Yashima, K. Maeda, T. Nishimura, Detection and amplification of chirality by helical polymers. Chem. Eur. J. **10**, 42–51 (2004)

107. A.J. Wilson, M. Masuda, R.P. Sijbesma, E.W. Meijer, Chiral amplification in the transcription of supramolecular helicity into a polymer backbone. Angew. Chem. Int. Ed. **44**, 2275–2279 (2005)

108. C. Zhao, K. Ouyang, J. Zhang, N. Yang, Synthesis and properties of optically active helical polymers from (S)-3-functional-3'-vinyl-BINOL derivatives. RSC Adv. **6**, 41103–41107 (2016)

109. P.K. Dutta, P. Jain, P. Sen, R. Trivedi, P.K. Sen, J. Dutta, Synthesis and characterization of a novel polyazomethine ether for NLO application. Eur. Poly. J. **39**, 1007–1011 (2003)

110. O.G. Morales-Saavedra, Non linear optical properties of novel amphiphilic azo-polymers bearing well defined oligo (ethylene glycol) spacers. Rev. Mex. Fis. **56**, 449–455 (2010)

111. D. Yan, S. Jing, Photoinduced reorientation process and nonlinear optical properties of Ag nanoparticle doped azo polymer films. Chin. Phys. Lett. **27**, 024204 (2010)

112. Z. Li, W. Wu, P. Hu, X. Wu, G. Yu, Y. Liu, C. Ye, Z. Li, J. Qin, Click modification of azo-containing polyurethanes through polymer reaction: convenient, adjustable structure and enhanced nonlinear optical properties. Dyes Pigm. **81**, 264–272 (2009)

113. M. Siwy, B. Jarzabek, K. Switkowski, B. Pura, E. Schab-Balcerzak, Novel poly(esterimide)s containing a push-pull type azobenzene moiety-synthesis, characterization and optical properties. Poly. J. **40**, 813–824 (2008)

114. N.K. Viswanathan, D.Y. Kim, S. Bian, J. Williams, W. Liu, L. Li, L. Samuelson, J. Kumar, K.S. Tripathy, Surface relief structures on azo polymer films. J. Mater. Chem. **9**, 1941–1955 (1999)

115. C.M. Gonzalez-Henriquez, C.A. Terraza, M. Sarabia, Theoretical and experimental vibrational spectroscopic investigation of two R_1R_2-Diphenylsilyl containing monomers and their optically active derivative polymer. J. Phys. Chem. A **118**, 1175–1184 (2014)

116. H. Kikuchi, H. Hanawa, Y. Honda, Development of polyamide-imide/silica nanocomposite enameled wire. Electron. Comm. Jpn. **96**, 41–48 (2013)

117. E. Grabiec, M. Kurcok, E. Schab-Balcerzak, Poly(amide imides) and poly(ether imides) containing 1,3,4-oxadiazole or pyridine rings: characterizations and optical properties. J. Phys. Chem. A **113**, 1481–1488 (2009)

118. S. You, C. Tang, C. Yu, X. Wang, J. Zhang, J. Han, Y. Gan, N. Ren, Forward osmosis with a novel thin-film inorganic membrane. Environ. Sci. Technol. **47**, 8733–8742 (2013)
119. S. Mallakpour, M. Zarei, Novel, thermally stable and chiral poly(amide-imide)s derived from a new diamine containing pyridine ring and various amino acid-based diacids fabrication and characterization. High Perform. Polym. **25**, 245–253 (2013)
120. A. Natansohn, P. Rochon, Photoinduced motions in azo-containing polymers. Chem. Rev. **102**, 4139–4175 (2002)
121. E. Schab-Balcerzak, A. Sobolewska, A. Miniewicz, J. Jurusik, B. Jarzabek, Photoinduced holographic gratings in azobenzene functionalized poly (amideimide)s. Poly. J. **39**, 659–669 (2007)
122. S. Hernandez-Ainsa, R. Alcala, J. Barbera, M. Marcos, C. Sanchez, J.L. Serrano, Ionic azo-codendrimers: influence of the acids contents in the liquid crystalline properties and the photoinduced optical anisotropy. Europ. Poly. J. **47**, 311–318 (2011)
123. J. Mysliwiec, M. Czajkowski, A. Miniewicz, S. Bartkiewicz, A. Kochalska, L. Polakova, Z. Sedlakova, S. Nespurek, Dynamics of photoinduced motions in azobenzene grafted polybutadienes. Opt. Mat. **33**, 1398–1404 (2011)
124. K.G. Yager, C.J. Barrett, All-optical patterning of azo polymer films. Curr. Opin. Solid State Mater. Sci. **5**, 487–494 (2001)
125. K. Akagi, T. Mori, Helical polyacetylene—origins and synthesis. Chem. Rec. **8**, 395–406 (2008)
126. H. Lu, J. Mack, T. Nyokong, N. Kobayashi, Z. Shen, Optically active BODIPYs. Coord. Chem. Rev. **318**, 1–15 (2016)
127. L. M. S. Takata, H. Iida, K. Shimomura, K. Hayashi, A. A. D. Santos, E. Yashima, Helical poly(phenylacetylene) bearing chiral and achiral imidazolidinone-based pendants that catalyze asymmetric reactions due to catalytically active achiral pendants assisted by macromolecular helicity. Macromol. Rapid Commun. **36**, 2047–2054 (2015)
128. S. Kumar, D.K. Tiwari, P.K. Dutta, J. Koh, Preparation and circular dichroism properties of chitosan/methoxycinnamaldehyde. J. Polym. Mater. **29**, 309–316 (2012)
129. J. Singh, S. Kumar, P.K. Dutta, Preparation and chiroptical properties of chitosan acid derivatives in dilute solution. J. Polym. Mater. **26**, 167–176 (2009)
130. H. Iida, S. Iwahana, T. Mizoguchi, E. Yashima, Correction to main-chain optically active riboflavin polymer for asymmetric catalysis and its vapochromic behavior. J. Am. Chem. Soc. **134**, 15103–15113 (2012)
131. D.Y. Wu, T.L. Zhang, Recent developments in linear chain clusters of low-valent platinum group metals. Prog. Chem. **16**, 911–917 (2004)
132. C. Bariain, I.R. Matias, C. Fdez-Valdivielso, C. Elosua, A. Luquin, J. Garrido, M. Laguna, Optical fibre sensors based on vapochromic gold complexes for environmental applications. Sens. Actuator B **108**, 535–541 (2005)
133. J.K. Bera, K.R. Dunbar, Chain compounds based on transition metal backbones: new life for an old topic. Angew. Chem. Int. Ed. **41**, 4453–4457 (2002)
134. S.C. Terrones, C.E. Aguado, C. Bariain, A.S. Carretero, I.R. Matias Maestro, A.F. Gutierrez, A. Luquin, J. Garrido, M. Laguna, Volatile-organic-compound optic fiber sensor using a gold-silver vapochromic complex. Opt. Eng. **45**, 040101–040107 (2006)
135. P. Rattanatraicharoen, K. Shintaku, K. Yamabuki, T. Oishi, K. Onimura, Synthesis and chiroptical properties of helical poly(phenylacetylene) bearing optically active chiral oxazoline pendants. Polymer **53**, 2567–2573 (2012)
136. M. Antonietti, K. Landfester, Polyreactions in miniemulsions. Prog. Poly. Sci. **27**, 689–757 (2002)
137. L. Ding, L. Chen, J. Deng, W. Yang, Optically active thermosensitive amphiphilic polymer brushes based on helical polyacetylene: preparation through "click" onto grafting method and self-assembly. Polym. Bull. **69**, 1023–1040 (2012)
138. T. Aoki, M. Muramatsu, T. Torii, K. Sanui, N. Ogata, Thermosensitive phase transition of an optically active polymer in aqueous milieu. Macromolecules **34**, 3118–3119 (2001)

139. G. Decher, J.B. Schlenoff (eds.), *Multilayer thin films–sequential assembly of nanocomposite materials* (Wiley-VCH, Weinheim, 2003)
140. J.H. Dai, A.M. Balachandra, J.I. Lee, M.L. Bruening, Controlling ion transport through multilayer polyelectrolyte membranes by derivatization with photolabile functional groups. Macromolecules **35**, 3164–3170 (2002)
141. H.H. Rmaile, J.B. Schlenoff, Optically active polyelectrolyte multilayers as membranes for chiral separations. J. Am. Chem. Soc. **125**, 6602–6603 (2003)
142. B. Zdyrko, M.K. Kinnan, G. Chumanov, I. Luzinov, Fabrication of optically active flexible polymer films with embedded chain-like arrays of silver nanoparticles. Chem. Commun. 1284–1286 (2008)
143. R. Yang, Y. He, Optically and non-optically excited thermography for composites: a review. Infr. Phy. Tech. **75**, 26–50 (2016)
144. B. Notario, J. Pinto, M.A. Rodriguez-Perez, Nanoporous polymeric materials: a new class of materials with enhanced properties. Prog. Mat. Sci. **78**, 93–139 (2016)
145. C.R. Mendonca, D.S. Correa, F. Marlow, T. Voss, P. Tayalia, E. Mazur, Three-dimensional fabrication of optically active microstructures containing an electroluminescent polymer. Appl. Phy. Lett. **95**, 113309–113313 (2009)
146. L. Angiolini, T. Benelli, L. Giorgini, E. Salatelli, R. Bozio, A. Dauru, D. Pedron, Synthesis, chiroptical properties and photoinduced linear birefringence of the homopolymer of (R)-3-methacryloyloxy-1-(4'-cyano-4-azobenzene)pyrrolidine and of the copolymers with the enantiomeric monomer. Europ. Poly. J. **41**, 2045–2054 (2005)
147. J. Gasiorowski, S. Boudiba, K. Hinger, C. Ulbricht, V. Fattori, F. Tinti, N. Camaioni, R. Menon, S. Schlager, L. Boudida, N.S. Sariciftci, D.A.M. Egbe, Anthracene containing conjugated polymer showing four optical transitions upon doping: a apectroscopic study. J. Poly. Sci. B Poly. Phy. **52**, 338–346 (2014)
148. A.J. Heeger, N.S. Sariciftci, E.B. Namdas, *Semiconducting and Metallic Polymers.* (Oxford University Press, Oxford, 2010) 978–0-19-852864-7
149. Y.J. Cheng, S.H. Yang, C.S. Hsu, Synthesis of conjugated polymers for organic solar cell applications. Chem. Rev. **109**, 5868–5923 (2009)
150. D.A.M. Egbe, H. Tillmann, E. Birckner, E. Klemm, Synthesis and properties of novel well-defined alternating PPE/PPV copolymers. Macromol. Chem. Phys. **202**, 2712–2726 (2001)
151. A.P. Vajpeyi, S. Tripathy, S.J. Chua, E.A. Fitzgerald, Investigation of optical properties of nanoporous GaN films. Physica E **28**, 141–149 (2005)
152. N. Maity, A. Mandal, A.K. Nandi, Synergistic interfacial effect of polymer stabilized graphene via noncovalent functionalization in poly(vinylidene fluoride) matrix yielding superior mechanical and electronic properties. Polymer **88**, 79–93 (2016)
153. C. Ren, Y. Chen, H. Zhang, J. Deng, Noncovalent chiral functionalization of graphene with optically active helical polymers. Macromol. Rapid Commun. **34**(17), 1368–1374 (2013)
154. A.C. Lopes, C.M. Costa, C.J. Tavares, I.C. Neves, S.L. Mendez, Nucleation of the electroactive g phase and enhancement of the optical transparency in low filler content poly (vinylidene)/clay nanocomposites. J. Phys. Chem. C **115**, 18076–18082 (2011)

Printed in the United States
By Bookmasters